Digital Magnetic Tape Recording for Computer Applications

MODERN ELECTRICAL STUDIES

A Series edited by Professor G. D. Sims

Head of Department of Electronics
University of Southampton

Digital Magnetic Tape Recording for Computer Applications

L. G. SEBESTYEN

Honeywell Information Systems Ltd.
Hemel Hempstead, England

LONDON

CHAPMAN AND HALL

First published 1973
by Chapman and Hall Ltd.
11 New Fetter Lane, London EC4P 4EE
© 1973 L. G. Sebestyen
Printed in Great Britain by
William Clowes & Sons Limited
London, Colchester and Beccles

SBN 412 11210 8

Distributed in the U.S.A.
by Halsted Press, a Division
of John Wiley & Sons, Inc.
New York
Library of Congress Catalog Card Number 73-6263

Preface

The purpose of this book is to provide the Information Scientist – and this term is used in the broadest sense to include the System Analyst, Programmer, Logic Designer, Computer Engineer – with a readable account of the principles and practices of data storage on magnetic tape. The emphasis is on physical and qualitative explanation of the phenomena and not on precise mathematical analysis.

There are a substantial number of excellent textbooks which cover all aspects of audio and instrumentation recorders, both theoretical and practical, but few are orientated towards the needs of the computer engineer. Computer textbooks which devote substantial space to ferrite core mainframe memories do not consider magnetic tape back-up stores as their province and usually restrict the subject to description of data formatting and organization on tape. On the other hand, magnetic recording handbooks discuss computer applications, at very best, in a short paragraph within the Digital Recording chapter.

The treatment concentrates on the magnetic aspects of digital recording and little space is devoted to mechanical problems of tape transportation, servo system design and digital electronics; these topics are adequately covered by many publications.

An extensive list of references is provided at the end of each chapter for the reader who wishes to pursue the subject in detail.

The introductory chapter indicates the place of digital magnetic tape recording in the hierarchy of computer memories, gives a first-level account of the process of storing data on magnetic tape, and compares the two main magnetic surface-storage families – tapes and discs – with each other. In Chapter 2 the various types of magnetizable materials which are currently used as storage mediums are described in some detail. Chapter 3 deals with the construction of magnetic heads. Although the principles of recording and reproduction are fairly well understood and theory permits the analysis and computation of the processes involved with the accuracy required by engineering practices, the physical insight into the recording mechanism is far from complete and much work is currently in progress. The complexity of the process defies the use of 'simple' models and explanations. Chapter 4 attempts to give an up-to-date account of the problems and results. The conversion of digital data into magnetized areas on the tape is the subject of Chapter 5. Various recording methods are compared and discussed with emphasis on the most widely used NRZ1 and PE coding systems.

A subject of considerable importance and often very confusing for the computer engineer is the theory and application of error detecting and error correcting codes in digital recording. An introduction to this subject is given in Chapter 6. Chapter 7 reviews the design philosophy of tape transport mechanisms.

As not only classic works on electricity and magnetism but many of the cited references use CGS system of units, Chapter 9 presents a CGS to MKS (SI system) conversion tables and recapitulates the basic relations between magnetic quantities. The practising engineer may find useful the list of standards on magnetic tape recording at the end of the book.

I am indebted to Mr J. Salmon, Director of Honeywell Information Systems for permission to publish the work and to Dr D. A. E. Whitehouse of the Department of Computer Science, The University of Manchester for the helpful criticism of the manuscript.

London   L. G. Sebestyen
November 1972

Contents

CONTENTS   ix

1 Introduction

1.1. COMPUTER MEMORIES

In 1840 Charles Babbage proposed an 'Analytical Engine' which consisted of a store, the mill (arithmetic unit) and a unit to control the sequence of operations. The store would have consisted of columns of wheels, each wheel being capable of resting in one of ten positions thus storing one decimal digit, and was to accommodate 1000 numbers of 50 decimal digit lengths. This machine like the earlier 'Difference Engine' was never completed but contained many of the basic ideas which were realized some 100 years later.

A characteristic feature of all modern digital computers is their ability to store both data and programs. The constants and variables on which the operations are to be carried out and the program, the sequence of instructions which specifies the operations, appear in identical format as a set of binary digits (bits). Many applications utilize the computer primarily as a device for storing large quantities of data; the numerical operations may be quite trivial. As any information, whether a number, a letter, or a program, appears inside the computer as a series of binary digits, the store or 'memory' of the computer must consist of a large number of two-state devices organized in such a way that information can be stored in the memory and retrieved from it at high speed. The capacity and speed of the memory are fundamental parameters of any digital computer. The early machines of the 1950's had storage capacities in the order of $5 . 10^4$ bits. The development of larger and more complex machines and systems necessitated the increase of the storage capacity and whereas systems in the 70's have as much as 10^7 to 10^8 bits random-access high-speed storage and over 10^{10} bits mass memory capacity, needs for 10^{12} bits start emerging [1]. The first 10^{12} bit laser mass memory has been reported to be operational at NASA Ames Research Centre in September 1972.

We used here several terms which need definition. A random-access store permits the retrieval of information which is held in it in any desired sequence; moreover the retrieval time is independent of the location of the information in the memory. In magnetic surface storage type mass memories – discs, drums and magnetic tape stores – the data retrieval is sequential, access time to any unit of information depends on its location in store. Current random access memories have an access time which is usually less than a microsecond whereas the access time for a disc-type store may be between 10 and 100 ms. average and as long as several minutes in the worst case for a magnetic tape unit. The 'unit of information' is a set of bits which are transferred, manipulated, and stored together. The length of this computer 'word' is the number of bits in the word. Many computers – particularly those types which are designed primarily for commercial applications – use 8-bit groups termed 'bytes' as the smallest unit; a word may be one or more bytes long. In random-access memories, a single word or byte is directly accessible; in bulk stores data are transferred between the central processor and the store in units of 'blocks'. The length of the block expresses the number of words or bytes in the block.

A large number of electro-mechanical, electronic, optical, and chemical phenomena are suitable for storing the two digits of the binary system. Basically, any device which has two stable, well-distinguishable states and can be changed over from one state to the other at fast rate and a very large number of times without deterioration of its performance, can be used as a memory element. The choice is dictated by such considerations as speed, cost per stored bit, size, power consumption and reliability. Optimum performance is obtained by judicious combination of devices which employ different technologies and organizations. Figure 1.1 compares the key parameters of the most important memory technologies. The fast random-access type mainframe memories can cost 100 to 500 times more per stored bit than the bulk memories, and above a certain size they tend to lose speed; hence the need for the much slower but cheaper bulk memories. A functional hierarchy* of memories is shown in Fig. 1.2; the main categories and areas of application are described in broad terms.

Since their introduction in the early 50's, ferrite core stores [2] have been universally adopted as the random-access mainframe memory standard. Other magnetic storage elements such as thin films and plated wire are used on much smaller scale. The magnetic memories are non-volatile: they retain the stored information independently of the power supply – for practical purposes – indefinitely. Semiconductor memories [3, 4, 5] started to erode the leading position of ferrite core stores in the early 70's. Registers (temporary stores) have been constructed of bistables from the earliest days of electronic computers. Integrated circuit technology has made large arrays of bistables an economic proposition. Metal-Oxide Silicon (MOS) devices have low cost per bit, small size and low power consumption but as the information is stored as a charge in a capacitor, they must be refreshed at a rate at which the charge leaks away. Both bipolar and MOS types are volatile: they lose the information if the power is removed. This necessitates the provision of an emergency power supply and a 'shut-down' routine which transfers the content of the semiconductor memory into a non-volatile store in case of power failure. Although semiconductor memories are typically small-capacity units, very large stores are becoming pratical propositions; 50 Mbit stores are now commercially available.

* Hierarchy: 'a graded organization'.

	Semiconductor	Magnetic core	Fixed head discs	Moving head discs	Magnetic tape	Optical Laser Holographic Photographic (non-erasable)
Typical cost per 1000 bits in $ cents (US)	600–1500	500–1000	Small 100–250 Large 50–100	7–10 7–8	Slow 1–3 Fast 3–6	Forecast 0.2–1.0 for very large capacities
Capacity in bits	Typically 1000 per unit, up to 50×10^6	Between 1 and 64 K per unit, up to 2 M	Small 0.1 10^6–5.10^6 Large 5.10^6–50.10^6	20.10^6–100.10^6 100.10^6–800.10^6	Slow 10.10^6–300.10^6 Fast 50.10^6–300.10^6	Up to 10^{12}
Access time μ s	0.2–0.6	0.3–0.6	Small 8.10^3–18.10^3 Large 6.10^3–18.10^3	20.10^3–120.10^3 20.10^3–100.10^3	5.10^3–5.10^9	
Transfer rate in 10^6 bytes/s	1–4	1–3	0.1–2.5	0.1–0.8	Slow 0.01–0.05 Fast 0.05–1.2	

Fig. 1.1. Comparison of key memory parameters.

Type	Application area
Bistables	The basic one-bit memory elements in the control and arithmetic units of the computer
Registers	Short-term one word length stores for instructions, interim results of computation, addresses, data
Scratch-Pad Memories	Temporary stores with capacity from one to a few hundred words for holding interim results of computations, subroutines, reference tables, etc.
Main Frame Memories	Random access stores with capacities usually in excess of 1000 words for storing data, programs, results of operations, tables, etc.
Bulk Stores	Sequential access stores with large capacities (10^5–10^{10} words) for storing data, programs, tables, etc.
Read-Only Memories	Non-erasable random-access stores for holding frequently used routines, microprograms, tables with capacities from one to a few hundred words

Fig. 1.2. Functional memory hierarchy.

Another broad category of stores is listed in Fig. 1.1 under the heading of Optical Memories. They are essentially non-erasable types, and although laser and photographic memories are used in special-purpose and experimental computers, they have not yet reached the stage where comparison on a commercial basis is meaningful.*

The purpose of this brief survey was to establish the position of magnetic tape stores in the computer memory hierarchy. They are low-cost, large capacity, sequential access devices with well established and proven technology. There is a continuous improvement in reliability, storage capacity per unit area, and cost per stored bit. Progress over the last two decades is more of the nature of refinement of technology than that of revolutionary innovations.

1.2. MAGNETIC TAPE STORES

The history of magnetic recording begins with Valdemar Poulsen's patent application for a 'Method of Recording Sounds or Signals' in 1899. Between 1900 and 1910 a further 11 patents were taken out [6, 15] and although some commercial exploitation of the ideas was attempted at the turn of the century, magnetic recording of sounds and telegraph signals remained, to a large extent, a scientific curiosity until the late 30's. Magnetic recorders using steel wire or steel tape were used in some specialized areas, but the real break-through came in the years immediately after the Second World War with the replacement of wire, as a recording medium, by a thin flexible layer of plastic tape coated with a magnetizable material. Since then analogue recording – where the prime objective is the faithful reproduction of the input waveform – has expanded at a dramatic rate, first in the field of sound recording, later in the areas which

* Reference [14] contains an extensive review on the state of the art.

require wider bandwidth such as scientific and industrial applications and video recording.

The first reports of use of magnetic recording techniques for storing digital data in conjunction with computers, seem to have been published in 1947-48 [7, 8]. From the 50's storage on a magnetic surface became the standard method for provision of large capacity memories for computers [9].

In simplest terms, data storage on a magnetic surface is effected by magnetizing a minute volume of a thin layer of the magnetizable material by a transducer head. The recording – or writing – process leaves behind a series of elementary permanent magnets; depending on the recording method which is used, it is either the presence or absence of these magnets in a particular area of the tape, or their polarity, which represent the logical zeros or logical ones.

Digital data recording techniques employ nearly exclusively longitudinal recording so only this method will be discussed and analysed in the forthcoming paragraphs. The longitudinal recording head (Fig. 1.3) consists of a ring-like

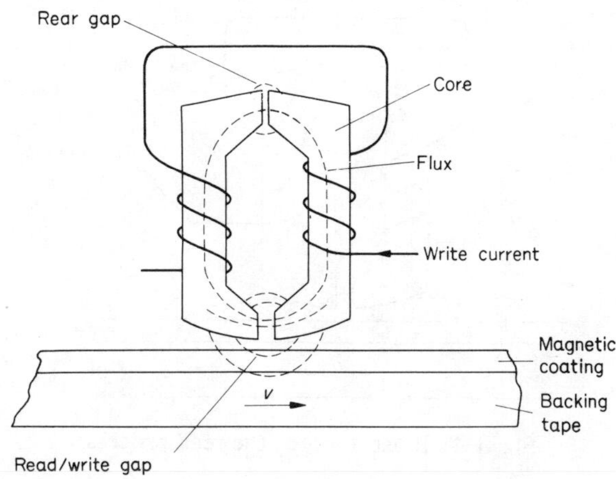

Fig. 1.3. Illustration of the write process.

structure of a high-permeability material with a very short air gap and a winding. The current in the winding generates a flux in the core and the 'fringing flux' from the air gap produces a minute permanent magnet in the magnetizable material which is brought in close contact with the head. The path of the write head along the surface of the medium is called a track. Data can be recorded serially on a single track or simultaneously on several parallel tracks.

The construction of the read head is similar to that of the write head, only the gap is shorter; the number of turns and track width may be different. The read head is placed in close proximity to the magnetized medium so that some of the flux generated by the elementary permanent magnets finds its way through the read head winding (Fig. 1.4). If the magnetized medium is moved across the head – mechanical considerations dictate to move the medium and not the head – a voltage is generated at the read head terminals. This voltage which represents a digit, is detected, amplified, shaped and eventually transferred into various registers of the computer.

The process as described up to now utilizes the same principles as the well-known and ubiquituous sound recorder. In sound and instrumentation recording, i.e. analogue recording, the objective is the faithful reproduction of the input signal. The medium and recording method is required to permit a wide dynamic range (ratio of highest to lowest undistorted signal levels). The reduction of background noise and achievement of good signal-to-noise ratios throughout the dynamic range is of fundamental significance. The medium is carefully erased with an AC signal and the inherent non-linearity is corrected by superimposing a high-frequency 'bias' on the signal to be recorded. Reduction of the reproduce amplitude level, or even its total loss for a few cycles, may go unnoticed and minor imperfections of the medium may be tolerated.

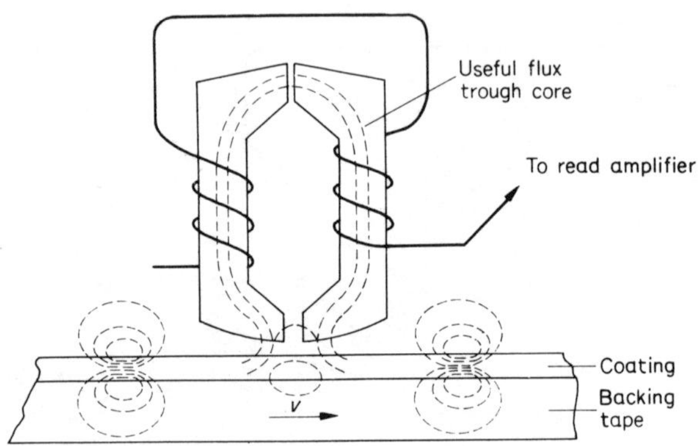

Fig. 1.4. Illustration of the read process.

Digital data are stored as two unique states of magnetization. As maximum differentiation between the two states is obtained by saturation of the medium into opposite directions, most coding systems use 'saturation-to-saturation' recording. There is no medium linearity requirement and bias is not used. In saturation recording, the new information erases the previously written data, hence there is no need for AC erasure, although on tape, for reasons to be discussed later, a DC erase process precedes the writing process and on discs the edges of the recorded tracks are 'trimmed' by DC erasure. As the distance between the two signal levels of a digital recording system is relatively large, background noise is much less troublesome; however as each recorded bit is individually detected, a loss of one single bit represents a non-permissible distortion of data. Flaws in the medium which can cause either loss of amplitude of the recorded signal, or generate a short-duration spike are not tolerated, and computer-grade tape is checked for errors much more rigorously than audio-grade tape. The imperfections of the recording system are compensated by redundancy recording or various ingenious error-detecting and correcting systems (see Chapter 6).

Drums, discs, tapes

The thin layer of magnetizable material which constitutes the storage medium can be applied to a variety of backing materials and the relative motion between the medium and magnetic transducer heads can be effected by various classes of electro-mechanical devices. Accounting machines use magnetic ledger cards where a strip of magnetic coating is carried by a paper-based material; cash dispensers are operated by a plastic card carrying a magnetic strip. The above devices store a relatively small amount of data; the large-capacity computer stores fall into the broad categories of drums, discs and tapes.

Although the principle of storing data by surface magnetization is the same for drums, discs and magnetic tape, the differences in the mechanical arrangements for handling the coated surface and consequently the data organization are substantially different. Before dealing with our main subject, the magnetic tape stores, it seems appropriate to review the similarities and differences between the main categories [9, 10].

The term 'drum' means, in this context, a rigid mechanical body in the shape of a cylinder with the magnetizable material — coating, for the sake of brevity — carried on the cylindrical surface. The drum is rotated around the axis of the cylinder at constant speed and the magnetic heads are mounted in close proximity to the coating. Data are recorded in 'tracks' around the circumference. Usually there is one head per track although two or more can be used. The relative speed between the cylinder and the heads is constant but the maintenance of very small and constant head-to-medium spacing which, as we will prove in Chapter 4, is one of the critical parameters determining the performance, represents, a major mechanical engineering problem. Earlier constructions which used 'in contact' recording were severely limited in respect to life and reliability of heads and recording surface. The storage capacity is proportional to the surface area of the cylinder, and access time to the stored information is inversely proportional to the speed at which the drum is rotated. The mechanical problems inherent in large-capacity drum designs have limited their application to some specialized fields; for general purpose and in particular data processing use, drums were superseded by discs and tapes.

Magnetic discs are thin circular plates coated on one or both sides with the magnetizable material. A set of such discs is mounted on a horizontal (or vertical) shaft with sufficient spacing between adjacent discs to allow a magnetic head to be inserted between them. There are two basic categories known as fixed (head-per-track) and removable (moving head) types.

Fixed head discs have usually one head per track. The 'tracks' are concentric rings reducing in length as we go from the outer track to the inner near the axis.

Consequently reproduced voltage and bit density vary from track-to-track. The disc is not removable from the drive mechanism and is often hermetically sealed.

Nearly all modern designs utilize the 'flying head' principle [16]; the head mounting permits the heads to ride on a cushion of air when the disc revolves at full speed, and they either retract or land on the surface when the disc motion stops. The head-to-surface distance is usually between 50 and 100 μ in. Average access times are between 10 and 50 ms.

When very large storage capacities are required with relatively short access time, moving head discs are an attractive configuration for the magnetic surface. A stack of discs – up to 10 or even more – are mounted on a spindle and rotated at high speed. There is usually one head per surface which can be placed over the position to be recorded, or from which information is retrieved, by a servo-mechanism of great accuracy. Data are recorded, as on the fixed-head discs, in concentric rings; relative speed, reproduce voltage and bit density vary from track to track. The access time is considerably longer (25–100 ms.) than on fixed-head discs because of the time required to position the head. The discs are removable and exchangeable between drives of compatible design. Current disc designs use a bit density of approximately 2000 bits per inch and up to 400 tracks per inch although densities as high as 4000 bits per inch are becoming commercially viable.

The utilization of flying heads permits the application of head materials and coatings which are not practical on magnetic tape systems where the heads are in permanent contact with the tape.* A particular problem associated with flying heads on discs is that of the 'head crash'. The delicate balance of forces which keeps the head floating above the surface can be disturbed by minute particles entering the system from the environment, causing the head to hit and damage the recording surface. Hence fixed-head discs are often hermetically sealed; a bottle of compressed inert gas may be provided to keep the pressure inside the recording chamber slightly above the atmospheric pressure. Other discs have complex air filters which prevent entry of impurities with the cooling air stream. Some of the latest designs use a coating on the recording surface which permits many thousands of 'landings' before the performance deteriorates.

The magnetic tape transport and head design problems are radically different from those of disc drives, moving-head actuators and flying heads. The tape consists of a long strip of flexible base material on which a thin layer of magnetizable material has been deposited. The tape transport mechanism must pull the tape across the magnetic heads at constant speed and at closely specified tension; it must be capable of starting and stopping the tape within a few milliseconds in rapid sequences in either direction. This is quite different from drums and discs which are continuously rotating. The multi-track magnetic tape heads are usually rigidly mounted on the main casting† and are in contact with

---

* At high tape speed a thin film of air is formed between the head and tape. See Section 3.6.

† Some tape transports retract the head during the fast rewind operation.

the tape. The alignment of the individual heads in a multi-track head, and the head wear caused by the abrasive oxide which is the most widely used tape coating material, are problems specific to tape as distinct from drum and discs stores. The materials and head constructions will be described in Chapters 2 and 3, respectively. As magnetic tape stores usually employ a full-width DC erase head, the problem of incomplete over-writing of the previously recorded data by the new information does not arise.

From the standpoint of magnetic considerations the number of tracks per unit surface area could be the same on magnetic tape as on drums or discs; however drums and discs provide a rigid surface, the heads can be scattered around the drum periphery or distributed over the disc surface in case of head per track construction. The use of tape necessitates the mounting of all magnetic heads in the same line across the width of the tape. The physical size of the heads mounted in one stack is a limiting factor in track density. Other considerations which dictate the choice of track width and number of tracks per inch of tape are discussed in Section 5.3.

As the number of tracks per unit area is much lower on tapes than on discs (the standardized format uses 18 tracks per inch on tape as opposed to 100 to 400 tracks per inch on discs) the adjacent-track crosstalk, the coupling of some fringing fields from adjacent tracks into the head reading a selected track, is much lower on tape than on disc recording systems.

Functional elements of a tape store

For the purpose of this discussion a magnetic tape back-up memory can be divided into the following main elements: [11, 12, 13]

(a) The recording medium – magnetic tape – and means of storage for the tape.
(b) Magnetic transducer heads.
(c) The tape transport which provides means for moving the tape at a constant, predetermined speed across the write/read heads.
(d) Electronic circuits for writing and reading the data, and for controlling the tape motion.

All the above elements can be housed in a single cabinet; the main mechanical units – motors, heads, tape guides, reel holders – are usually mounted on a rigid casting which in turn is fitted into a cabinet. The cabinet may hold more than one transport, and the electronics may be shared between several transports.

The magnetic recording medium must be capable of retaining a sequence of magnetized states but it should not be too difficult to erase the previously recorded information. Those requirements dictate the use of a magnetically moderately hard recording material.* On the other hand the recording head has to concentrate the recording flux into the smallest possible area and the read head should represent a magnetic short circuit to the flux emanating from the recorded surface. Both considerations require a magnetically soft high-permeability material.

* See Chapter 9 for definition of 'soft' and 'hard' magnetic materials.

Mechanical considerations resulted in providing the necessary relative motion between head and medium by moving the medium. There are a few exceptions such as rotating head scanners (where the tape is stationary and one or more heads scan repetitively the same area) and video recorders where the rotating head scans the slowly moving tape. The magnetic tape memories are linked to the central processor of the computer via the magnetic tape control unit. The control unit provides a temporary data buffer which synchronizes the data transfer between the central processor and the tape unit, interprets and executes the motion control instructions from the central processor to the tape transport, and reports back to the central processor the status of the tape unit. The controller may contain the error detection and correction logic (see Chapter 6). Commercial applications, or large data banks, often utilize a battery of magnetic tape stores, and controllers are designed to handle four, eight or even sixteen tape transports.

The four main functional areas into which we divided the magnetic tape back-up memory illustrate the variety of disciplines which are involved: the recording and reproduction is largely a problem of magnetics, coupled with precision mechanical engineering for the heads, and tape transport. Transport design is a combination of mechanical engineering with electrical and control engineering skills; finally, the electronic circuit and logic designer and the programmer join forces in creating the read/write amplifiers, control logic and data organization system.

Advantages and disadvantages of magnetic tape

The principal advantage of magnetic tape as a data storage medium is that it can store a very great deal of information in a small volume and at low cost per bit. A typical 2400 ft long, $\frac{1}{2}$ in wide reel of tape weighs about 2 lbs and has $10\frac{1}{2}$ in diameter. If it is recorded – as specified by one of the widely used standards – on nine parallel tracks with 1600 bits/inch density, it can hold about 4×10^8 bits. Not all of the bits represent useful data because one or more bits in each character are recorded for checking purposes and some characters are instructions and labels; further, recording is done in blocks with unrecorded areas – interblock gaps – between the blocks. But even allowing for these dilutions of capacity, one 2400 ft long reel of tape can hold as many as 40 million eight-bit characters. An idea of this storage capacity can be obtained by calculating the number of volumes of printed books which can be stored on one reel of tape. A 500 page book may contain about 200 000 words, with the average of six letters per word. One letter is represented by one character on the tape so that one printed word occupies 6/1600 in and the full volume about 750 in, thus one 2400 ft reel of tape can store up to 37 such volumes.*

Other important advantages of magnetic tape memories are:
(a) The magnetic tape can be erased and re-used a large number of times. Errors are easy to correct.

---

* 6250 bits/in density machines, which are in an advanced stage of development will increase the capacity of a reel by a factor of four.

(b) There is no processing required. The information is available and can be checked immediately after recording.

(c) The storage is permanent, non-volatile and does not need refreshing at regular intervals. If handled with proper care tapes deteriorate very little over long periods (ten or twenty years) and can be frequently replayed. It allows recording and storing information away from the computer, a mode of operation which is not practical with other forms of stores such as core stores, thin film, or semiconductor memories.

(d) High data transfer rates can be obtained. Practical tape speeds and densities vary from 200 bits/in and 10 ips speed to 1600 bits/inch density and 200 ips speed corresponding to a range from 2000 ch/s to 320 000 ch/s transfer rate.

(e) High reliability. The rate of undetected errors can be as low as one in 10^{12} bits under practical conditions.*

(f) The cost per stored bit is the lowest of any bulk memories (Fig. 1.1).

The main disadvantage of magnetic tape storage systems is the slow access compared with magnetic discs or random-access memories and the serial nature of the process. In addition to these inherent limitations there are many practical problems. The magnetic tape is sensitive to mechanical handling, heat, magnetic fields and dust. It needs a complex, precision mechanical drive which is often the least reliable component in the computer system. As the presence of a particle of a foreign matter, such as dust or loose magnetic oxide can cause an error, it is essential to operate the magnetic tape system in a 'clean' atmosphere, usually in an air-conditioned room. The erasibility is not only an advantage but a risk; the information can be inadvertently destroyed not only by exposure to external magnetic fields but by an operator error. The usual protection against accidental erasure is a mechanical interlock; the reel is fitted with a ring which prevents switching the machine to write mode as long as the ring is in place.

The storage capacity of the recording system per unit medium area – which is a key performance measure of a data storage system – can be obtained by multiplying the bit density per linear inch of tape with the number of tracks per inch. Thus at the current commercially employed highest density the storage capacity is 1600 bpi x 9 tracks/$\frac{1}{2}$ in = 28 800 bits/square in. Much higher densities were reported and obtained under laboratory conditions and 6250 bits/in density machines are expected to become the industry standard of the future. It is essential to observe at this point that the track density is primarily determined by mechanical tolerances of the head, transport and to some extent, of the medium, thus in the first instance, a problem of precision mechanical engineering. On the other hand bit density is primarily a problem of magnetic recording physics and technology.

* High performance digital recorders have elaborate error detection and correction logic built into the control electronics. When an error is detected, the logic makes several attempts to correct it and reports a 'solid error' to the central processor only if corrective measures failed. Hence errors can be classified as detected, recoverable errors, detected, unrecoverable errors and undetected errors.

The magnetic tape store is essentially a serial-access device; the central processor must progress serially through the blocks of data until the required item is found. This limitation can be circumvented to some extent by programming techniques. Often the information to be stored is already ordered, or can be ordered, by simple means before feeding it to magnetic tape. In a typical commercial application such as storing and updating a Sales Ledger a master tape would be prepared containing – among other data – the customer's name, address, account reference and opening balance. Next, movement data are written on a second tape probably from punched cards prepared from the source document. The two tapes are then linked on-line to the central processor. Data are read into the store from the master tape for the first account on the list and then the movement data relating to the same account from the second tape. The master data and the movement data are merged calculating the new balance and written on a third tape which now becomes the new master. This organization removes much of the searching and waiting time caused by the serial nature of the storage system.

1.3. DATA ORGANIZATION ON MAGNETIC TAPE

Magnetic tape storage systems allow the interchange of data and programs between various computer installations only if the recorded formats and some critical parameters of the tape transports, heads and electronics conform to the same standards. From the large variety of recording formats which were incompatible, the convention used by International Business Machines has emerged as a *de facto* standard. Eventually various national and international bodies such as the American National Standards Institute, European Computer Manufacturers Association, British Standards Institute, International Standards Organization have issued recommendations and standards for all critical characteristics of the magnetic tape, heads, recording methods, data organization, error checking, etc. which have now been widely adopted by all major computer and magnetic tape equipment manufacturers. The full discussion of the standards and their implications are postponed until the principles of writing and reading on magnetic tapes are reviewed. However it is essential to introduce, at this stage, some of the terms and conventions of the data organization. There is considerable variation on terminology between textbooks, manufacturers, and British and American usage; we will use throughout this book the definitions of British Standards B.S. 4503: 1969 and B.S. 4732: 1971. The general format of a recorded tape is shown in Figs. 1.5 and 1.6.

The standardized tape width is $\frac{1}{2}$ in (although older machines use $\frac{3}{4}$ in and 1 in wide tape) and data are recorded or seven or nine tracks in parallel. The seven or nine data bits (including a redundant parity check bit) are recorded in one row (frame). A number of the characters recorded on the tape will form a *block*. The blocks are separated from each other by an *interblock gap*. The block is the smallest unit of data which is transferred between the *tape transport* and the

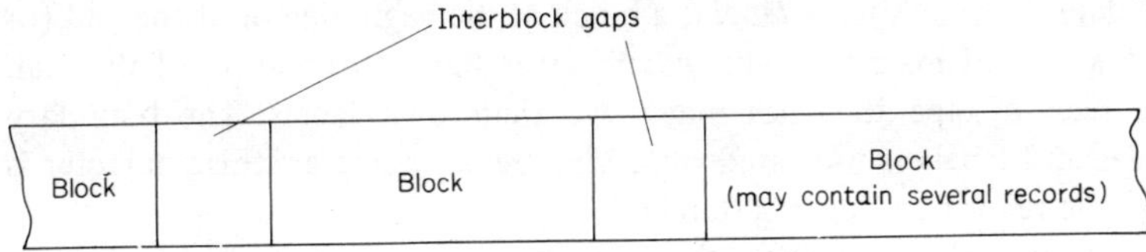

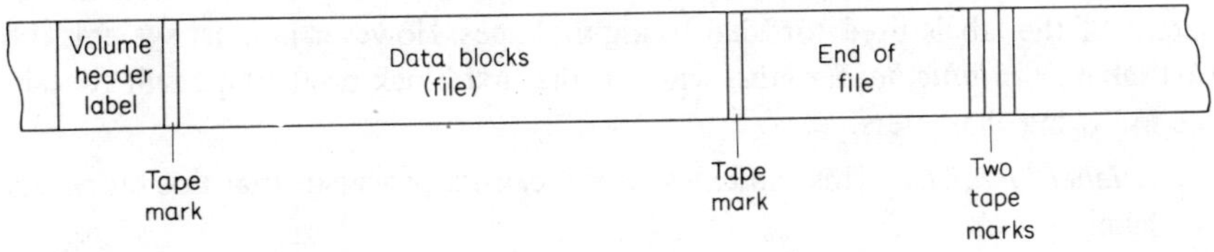

Fig. 1.5. Data organization on magnetic tape. Blocks, records, files.

central processor. The tape transport comes to a halt in the interblock gap and waits for further commands from the central processor. The length of the blocks may either be constant or variable; this is a matter for the computer data

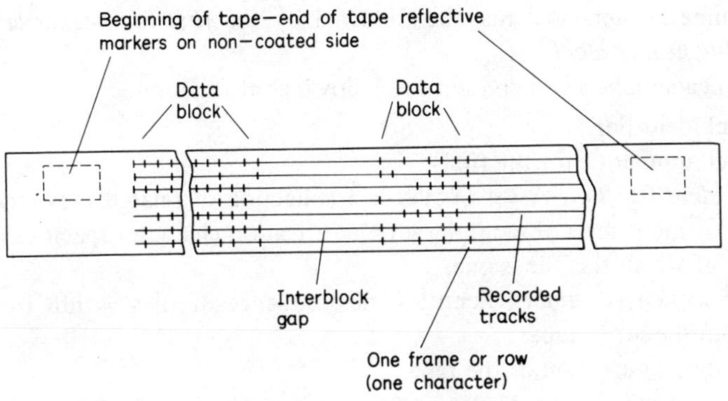

Fig. 1.6. Bit-parallel character-series recorded tape.

organization but the standards specify a minimum and maximum block length. A block may contain one or more records. A record is 'a collection of related items of data which for operating system logic purposes is treated as a unit of information'. Conceptually, a record corresponds (in the context of data processing applications) to a transaction, a customer's account, etc. In other contexts the delineation of a record may be arbitrary and is determined by the designer of the information formats. A *file* - quoting B.S. 4732: 71 - is a major collection of data consisting of all the records pertaining to a subject. In the context of data processing it relates to such collections as payroll file, an inventory file, etc. In other contexts the delineation is arbitrary and may relate to a set of data, a table, etc. A *label* is a block at the beginning or at the end (or at both) of a reel of magnetic tape which serves the identification of the reel. Within the reel of tape the files may have their own labels. The boundary between files and labels is indicated by a *tape mark* character; this character is specified by the relevant recording standards.

The beginning of the tape where the recording may start is marked by a small reflecting strip which is detected by a mechanism on the tape transport (*load point marker*). Similarly the end of the usable area of the tape is marked by the *end of tape* reflecting strip (Fig. 1.6). There is considerable variation in the details of the labels used for identifying the tapes. However in most systems the first label or *volume header label* which is the first block on the tape will contain the following characters:

(a) *A label identifier.* This indicates to the central processor that this block is a label.
(b) A serial number allocated to the tape. This is a permanent number identifying the tape and does not change with the data recorded on it. (This rule is not necessarily followed at every computer installation.)
(c) Various characters in the label may specify whether access to the contents of the tape is restricted or unlimited. The restrictions must be specified by the user.

The volume header label may be followed by the *first file header label* and a *second file header label.*

The file header labels will contain the following information:

(a) Label identifier.
(b) Label number (1 for the first).
(c) *Set identification.* A set of files is a collection of related files recorded on one or more reels of tape; the set identification character specifies the set of files of which this file is one.
(d) *File sequence number* denoting the sequence of files within the tape or within the set of tapes.
(e) The date of creation of the tape.
(f) The expiration date. If the current date is equal or later than the expiry date the tape may be overwritten.

The files are terminated by an end of file label which contains − in addition to the same characters as the header label − a *block count number.* The block count is checked against a count made by the central processor of the blocks which were processed to show whether any of them have been missed. The last block on the tape is the end of volume label. Some conventions use the term *field* to describe a group of characteristics, a number of fields a record and a number of records a block. The B.S. 4732: 71 standard does not specify explicit indication of the boundaries between records; others have *end of field, end of record* and *end of block* marks.

Magnetic tapes recorded on various installations must be 'hardware compatible' to allow interchangeability. By this we mean that such factors as tape width, track width, recording formats, densities, interblock gap lengths, etc. must be identical. Other factors, in particular the labelling system, may have differences; the tape recording system may still be able to handle the tape but the computer has to interpret the information.

REFERENCES

1. Weil, J. W. (1971), 'An introduction to massive stores', *Honeywell Computer Journal*, **5**, 2, 88–92.
2. Rickards, B. W. (1970), 'Core memories: the perennial leader', *IEEE Trans. Mag.*, **Mag-6**, 791–95.
3. Eimbinder, J. (Editor) (1971), *Semiconductor memories*, Wiley-Interscience.
4. Ayling, J. K. and Moore, R. D. (1971), 'Main monolithic memory', *IEEE J. Solid State Circuits*, **SC-6**, 276–279.
5. Jordan, W. F. (1971), 'Main memory: past, present and future', *Honeywell Computer Journal*, **5**, 2, 52–57.
6. Wilson, C. F. (1956), 'Magnetic recording 1888–1952', *Ire Trans. Audio*, **Au-4**, 3, 53–81.
7. Engineering Research Associates Inc., St. Paul, Minn., 'Storage of numbers on magnetic tapes', Report No. PB 99667, June (1947).
8. West, C. F. and De Turk, J. E. (1948), 'A digital computer for scientific applications', *Proc. IRE*, **36**, 12, 1452.
9. Richards, R. K. (1957), *Digital components and circuits*, D. Van Nostrand Co. Inc., 263–396.
10. Hess, H. (1963), 'A comparison of discs and tapes', *Comm. ACM*, **6**, 10, 634–663.
11. Franklin, D. P. (1960), 'Factors influencing the application of magnetic recording to digital computers', *J. Brit. IRE*, **20**, 9–21.
12. Edwards, D. G. B., Aspinall, D. and Kilburn, T. (1964), 'Design principles of magnetic tape system for the Atlas computer', *The Radio and Electronic Engineer*, **27**, 65–73.
13. Willis, D. W. and Skinner, P. (1960), 'Some engineering aspects of magnetic tape system design', *J. Brit. IRE*, **20**, 867–876.
14. Thompson, S. A. (Editor) (1971), 'Memories course', *The Electronic Engineer* (USA).
15. John Crerar Library, Bibliography Series No. 1 (1950), *Magnetic Recording 1900–1949*.
16. Brooke, R. and Mo, F. (1968). 'Flying heads for disk drives', *Honeywell Computer Journal*, **2**, 3, 34–45.
17. Matick, R. E. (1972), 'Review of current proposed technologies for mass storage systems', *Proc. IEEE*, **60**, 3, 266–289.
18. Hoagland, A. S. (1972), 'Mass storage: past, present and future', *AFIPS Conference Proceedings*, **41**, 2, 985–991.

2 Magnetic recording materials; the magnetic tape

In this chapter we wish to review the composition, preparation and properties of recording media. The emphasis will be on iron oxides which are currently the most widely used materials.

The recording materials can be classified into broad categories of 'particulate media' and 'continuous' metallic media. The magnetic materials which consist of fine particles can be divided further into classes of oxides (such as gamma-Fe_2O_3, Fe_3O_4, CrO_2) and metal particles (iron, cobalt, nickel). The continuous metal films usually contain Co–Ni–P, Fe–Ni–Co or Co–P combinations with various amounts of the constituents. The early recorders of Poulsen [1] and the dictating machines built around the turn of the century used thin steel wire as the recording medium. It is interesting to note, that the idea of preparing metal film recording surfaces by electrochemical deposition was patented by Pedersen in 1906. However, steel wire and steel tape remained the basic recording material to the 30's [2] and are still used in some special areas. Homogeneous metal strips have high noise level, poor resolution, are difficult to splice, and do not conform to the read/write heads. The mechanical and magnetic properties cannot be varied independently.

A major breakthrough came with the separation of the magnetizable material and a non-magnetic carrier (backing material). This arrangement permitted variation of the thickness and magnetic properties of the recording medium independently of the carrier. The first experiments around 1928 used paper tape which was replaced by cellulose acetate, polyvinyl chloride and eventually by polyesters.

The theory of magnetic recording which will be discussed in some detail in Chapter 4, gives an indication of the requirements for the magnetic parameters of a good recording medium. The material shall have a high coercive force to provide stability against self-demagnetization and, as in saturation digital recording the induced voltage in the reproduce head is proportional to the residual intensity of magnetization, high remanence is desirable. Further, the hysteresis loop shall be 'square', i.e. the ratio of B_r to B_s shall be high (Fig. 2.1).

The permeability does not seem to be of prime significance. Some further requirements limit the choice of materials which satisfy the coercivity and remanence criteria: it must be possible to form the material into a very thin, very smooth, and if used on tape, flexible layer. The elementary magnetic units, domains, must be small to provide satisfactory short-wavelength resolution and low noise. The material must be insensitive to normal variation of temperature and humidity and retain its magnetized state over very long – 20 years – periods. It should be reproducible to a very high degree by mass production with low raw material and manufacturing costs.

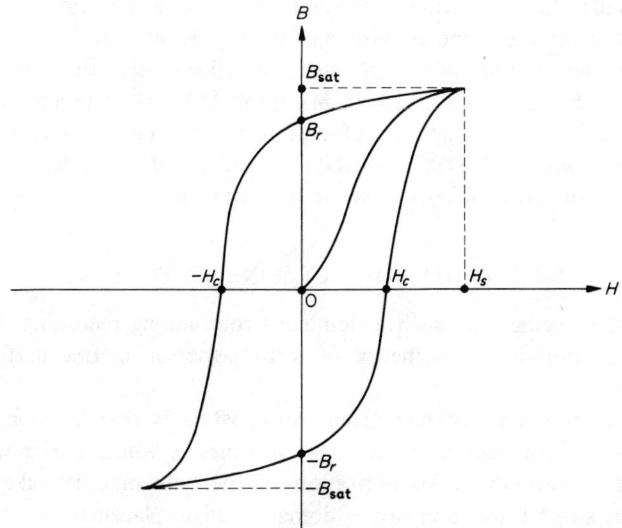

Fig. 2.1. Hysteresis loop.

The material which satisfies many of the basic requirements and for which manufacturing techniques and technologies have been perfected during a period of some 20 years is gamma-Fe_2O_3. Systematic investigation of over a thousand different materials has produced but few rivals. The overwhelming majority of both analogue and digital magnetic tape recorders uses this iron oxide as recording medium. The situation is slightly different on discs and drums where the physical construction permits the application of thin metal films, but oxide coating is still extensively used.

The relation between the recording performance, in particular the reproduce output voltage, pulse width, resolution, noise and the magnetic properties of a given material, are highly complex [3, 12, 25, 39]. Whereas the previously stated requirements for high coercivity and remanence are, in general valid: the effect of both factors appears to depend on the composition and thickness of the medium (i.e. whether particulate material with binder or metal film). Typical oxide coatings are 100 to 600 μ in thick, whereas uniform metal films can be produced with less than 5 μ in thickness. Further, oxide coatings permit only small variation in coercivity (between 200 and 350 Oe); on metal films the range can be as wide as 20:1. The advent of thin metal films necessitated the revision of the previous theories and triggered a series of both theoretical and experimental studies which gave some conflicting results [3, 4, 5, 6, 7, 8, 9, 10, 11] in respect of the effect of magnetic parameters on recording performance.

Because of the interaction of magnetic and mechanical parameters such as read and write head gap length, surface roughness, head-to-medium separation, binder composition in particulate media, write current, medium thickness, the experimentalist faces a difficult problem in isolating the effects of magnetic parameters on the width and amplitude of the reproduce pulse.

Much of the material in this chapter is based on publications of C. D. Mee [13, 14], D. E. Speliotis [15], J. C. Mallinson [16], G. Bate and J. K. Alstad [17]. Practical details of tape manufacturing and testing are described by D. F. Eldridge [18] and E. D. Daniel and D. F. Eldridge [19]. A lucid explanation of fundamentals of ferromagnetism can be found in the book of A. Morrish [20].

2.2. PROPERTIES OF FINE PARTICLES

As particulate media play such a dominant role among recording materials, a brief introduction into the theory of ferromagnetism in fine particles seems appropriate.

The classical theory of ferromagnetism postulates that ferromagnetic substances are divided into domains – Weiss zones – which are spontaneously magnetized to saturation. Magnetization of the bulk material takes place by variation in size of the domains – domain wall displacement – and through change of direction of magnetization by rotation of spontaneous magnetization. The magnitude of this spontaneous magnetization shows a temperature-dependence; the general shape is that of a monotonous decrease with increasing temperature reaching zero at the Curie point of the material.

In single-domain particles the primary mechanism is domain vector rotation. The intrinsic coercivity of such fine particle materials can be derived from magnetocrystalline and shape anisotropy. Magnetocrystalline anisotropy arises because of the existence of a preferred direction of magnetization with respect of the crystal axes. Shape anisotropy arises in particles which are non-spherical because in order to minimize the magnetostatic energy associated with the demagnetizing field within the particle, the magnetization must be parallel with the longest axis. Finally, the interaction between the magnetization of a particle and the mechanical stress acting on it determines a stress anisotropy. The latter has no application in tape materials. The actual coercivity is obtained after mixing the ferromagnetic material with a non-magnetic binder, usually in $\frac{1}{3}$ oxide – $\frac{2}{3}$ binder volume ratio so that typical computer tapes exhibit 250 to 300 Oe coercivities. The theory postulates a linear addition of the shape and crystal anisotropy terms of coercivity and leads to the conclusion, that approximately 30% of the observed coercivity is the consequence of crystal anisotropy, the rest, of shape anisotropy. Thus shape anisotropy is the dominant factor which determines the coercivity of the particulate media [16, 17, 19].

2.3. EFFECT OF PARTICLE SIZE

It has been implied in the preliminary discussions that it is desirable for the recording material to exhibit a near-rectangular hysteresis loop, high saturation flux density, low temperature dependence, and sufficiently high coercivity to prevent self-demagnetization. Other considerations made it desirable to choose single-domain particle sizes. The coercivity was shown to be strongly dependent on particle shape. The critical dimension beyond which a particle ceases to be single-domain, depends on material constants with ranges from 100 Å for low-anisotropy high-magnetization materials such as iron, to several thousand Angstroms for highly anisotropic low-magnetization oxides. Increasing particle size – exceeding the single-domain particle size for the material – causes a strong decrease in coercive force. At the other hand, minimum desirable particle size is determined by the onset of the phenomenon called superparamagnetism. If the particle size is less than a critical value determined by material constants and particle shape, the coercive force becomes zero and magnetization becomes a single-valued function of time, temperature and applied field. Superparamagnetic behaviour occurs when the magnetic energy of a particle is of the same order of magnitude as the thermal energy kT. In other words, as the volume of material is reduced, a condition is reached where the randomizing forces due to thermal energy are stronger than the alignment forces. The calculated value for acicular* γ-Fe_2O_3 with 5 : 1 length/width ratio gives 750 Å/150 Å length/width dimensions at which superparamagnetism will occur. For spherical γ-Fe_2O_3 the corresponding radius is 350 Å [16].

Consequently, we have now defined the desirable particle size: it shall be above the onset of superparamagnetism but not exceed the single-domain size. The theoretical model of single-domain magnetization predicts a 0.5 ratio of remanence to saturation for the case when shape anisotropy is dominant. However, with uniaxial anisotropy the 'squareness ratio' can be improved by aligning the particles in the direction of the applied field and the ratio can be increased to be in the range of 0.65 to 0.85.

The temperature dependence of coercivity and remanence in particulate materials is a function of material constants, shape and particle size. As long as the shape anisotropy is dominant and the range of operation is sufficiently remote from the Curie point of the material, the temperature dependence is not significant. Typically, for acicular γ-Fe_2O_3 which has a coercivity of 300 Oe at room temperature, the change is less than 5% in the 0–100°C range.

* Needle-shaped.

2.4. THE GAMMA Fe_2O_3

The material most widely used in general purpose particulate coatings on tape is $\gamma\text{-}Fe_2O_3$. The prefix 'gamma' distinguishes the ferromagnetic form from the non-ferromagnetic alpha Fe_2O_3.

As the weight of a sample of particulate material can be measured much more easily and to higher degree of accuracy than the volume or cross-sectional area, it is customary to express the magnetic properties of material in terms of saturation moment per gram, σ_s. The saturation flux density B_s is then calculated as $4\pi\sigma_s\rho$ where ρ is the density of the material. The particles of $\gamma\text{-}Fe_2O_3$ used in recording surfaces usually have lengths of 0.5 to 1.0 μm and length-to-width ratios between 7 : 1 and 5 : 1. Other typical parameters are σ_s = 76 EMU/g, ρ = 5 g/cm^3, B_s = 4700 G, H_c = 200–300 Oe. As gamma Fe_2O_3 is rapidly converted to $\alpha\text{-}Fe_2O_3$ at temperatures above 500°C, it is difficult to establish the Curie temperature; different methods yielded values between 470°C and 647°C. The accepted most likely value is 590°C [17].

The most common process for making the particles uses hydrated ferric oxide [$\alpha(FeO)OH$] which exists in needle-shaped crystalline form or can be made by a process described by M. Camras [38]. After dehydration red acicular crystals of hematit, $\alpha\text{-}Fe_2O_3$ are obtained. The hematit crystals are reduced by heating in hydrogen at 300°–400°C to black crystals of ferrosoferric oxide (Fe_3O_4 = $FeOFe_2O_3$ = magnetite). Finally reoxidation of the magnetite at lower temperature (250°C) produces the acicular $\gamma\text{-}Fe_2O_3$.

2.5. OTHER OXIDE MATERIALS

The use of Fe_3O_4 particles, although its saturation point and coercivity is higher than that of γFe_2O_3, is limited, apparently, for reasons of higher temperature dependence of coercivity, higher print-through and suspect long-term stability of storage. As adjacent layers of magnetic tape when wound on a reel are only separated from each other by the 1 to 1.5 mil thick backing material, each layer experiences the magnetic field of its neighbours. As the adjacent fields are much smaller than the writing fields, the material will be resistent to print-through if this low field does not leave behind any remanent magnetization. The print-through effect may be disturbing in high quality audio recording where adjacent layers can produce background noise and low-level signals on an erased part of the tape.

Other oxide materials which were considered are cobalt-doped iron oxide, and chromium dioxide. Cobalt increases the room temperature coercive force at nearly linear rate with the cobalt content and cobalt-substituted $\gamma\text{-}Fe_2O_3$ tapes which have coercivity of 500 Oe have been commercially available for some time under 'high energy' tapes designation. The main disadvantage of cobalt addition seems to be increased temperature sensitivity over $\gamma\text{-}Fe_2O_3$ material, higher print-through, and stress-induced instability, connected with magnetostrictive properties of cobalt. Other ferrites – Mr, Ba, Sr – which have been investigated exhibit excessive temperature dependence of coercivity.

A non-ferrous material which has the properties required for recording media is chromium dioxide (CrO_2). Samples with 5 : 1 length/width ratio exhibited 250 to 350 Oe coercivity at normal packing ratios. The saturation magnetization is 106 EMU/g. It has a low Curie point (119°C). Chromium dioxide tapes are manufactured in commercial quantities for both audio and digital recording but their application is not widespread [17, 47].

2.6. METAL PARTICLES

Metal particles have higher saturation magnetization than oxide particles. This would allow the use of thinner recording layers without loss of output. The high-density performance is helped by high coercivities obtainable with metallic single-domain particles. The materials which are of prime interest are iron-cobalt-nickel alloys. Preparation of metal particle surfaces yielding coercivities above 700 Oe and remanent intensity of magnetization of 222–238 EMU/cm^3 were reported (compared with the 120 EMU/cm^3 for γ-Fe_2O_3 at the same particle packing density).

As metal particles could be made smaller without running into problems of superparamagnetism, a better noise characteristic is expected. A further practical advantage of metal particle tapes is that their surface is less abrasive than that of oxide-coated tapes thus reducing the wear on the magnetic head. In view of the advantages of metal particles over oxide particles it is somewhat surprising that metal-particle media are not more widely used. The problems appear in large quantity production of particles with carefully controlled and appropriate uniform magnetic properties. Another problem is that fine metal powders are liable to burn spontaneously on exposure to atmosphere which necessitates the use of plastic coating or other preventive measures.

2.7. METAL FILMS

Continuous metal film surfaces allow a wide range of coercivities (50–1500 Oe) together with high saturation levels. The coating may be considerably thinner than that of oxides and still provide suitable output levels. The materials of prime interest are Co, Ni, P and their various combinations. The surfaces are prepared by electrochemical methods or vacuum deposition.

Although thin-metal film recording media have been in use on drums and discs for some considerable time, they are not normally used on tapes. The problems seem to be connected with poor wearing characteristic (on discs the head floats above the surface and is not in contact with the media), sensitivity to 'atmospheric' conditions in particular humidity and high temperature which may cause deterioration of unprotected surfaces, poor adhesion to plastic tapes and difficulty in producing large quantities of carefully controlled dimensionally and magnetically uniform surfaces. Finally, one reason which may be a contributing factor to the reluctance of accepting metal film tapes is that they are incompatible with existing machines.

2.8. BASE MATERIALS AND BINDERS

The most widely used base material for computer-grade magnetic tape is polyester (oriented polyethylene terephtalate) which has, to a large extent, superseded the cellulose acetate and polyvinyl chloride materials used in earlier and in audio-type applications. Polyesters have high dimensional stability, good ageing properties, low hygroscopic coefficient of expansion (thus moisture-induced distortion is minimal), high tensile strength and exhibit high resistance to organic solvents. Electrical resistance is very high (typically 10^{18} ohm cm) which may lead to static charge accumulation problems.

The essential ingredients of the binders are plastic resins soluble in organic solvents. The resins are basically two types, thermoplastic or thermosetting. In thermoplastic binders the solvent is driven off after coating and no polymeric change takes place; the coating remains soluble in the same solvents which were used during preparation. In the thermosetting binders the polymer chain grows under the influence of the catalyst and heat treatment and they are insoluble in organic solvents. The actual binder composition which is suitable for a particular oxide and application is the result of extensive research over a long period and is proprietary information. The main components are vinyls, epoxy, polyester, polyurethane or polycarbonate resins, with additives such as finely dispersed carbon particles for the purpose of reducing surface resistance, agents which influence viscosity and prevent agglomeration during the preparation process, lubricants such as silicones and fungicides to prevent fungus growth [32].

2.9. OUTLINE OF THE MANUFACTURING PROCESS

The preparation of coating usually starts with the mixing of the constituents, the oxide, resin and solvents. As the small magnetic particles have a tendency to agglomerate, a very thorough mixing is essential. This is often undertaken in two steps, first preparing a near-homogeneous pre-mix, which is then followed by the dispersing procedure. The mixing vessel contains steel or ceramic balls and is rotated, or the vessel is stationary and the balls are stirred by a rotating axle. The size of balls, rotating speed, and dispersion time are carefully controlled and are important factors in determining the final properties of the coating. After the mix has reached the required stage it is applied uniformly to the base material. Control of uniform thickness is of prime importance. The coating thickness is determined either by a 'knife', a precision steel blade positioned above the continuously moving base material onto which the mix is coated from the applicator, or by a process similar to the one used in some types of printing presses. The surface of the gravure roll carries a series of closely spaced fine grooves. The coating material is applied first to the gravure roll and fills the grooves, with the surplus carefully removed. The quantity of mixture is now precisely determined by the groove capacity of the gravure roll. The gravure roll deposits the mixture onto the continuously moving base material and the mixture is spread out by a smoothing roller [18, 19, 32].

The coating is followed by an orienting and drying process. The coating process itself tends to introduce some degree of particle orientation. The magnetic field which is applied to the still viscous coating causes the acicular particles to align their longer axes with the magnetic field. The direction of the field is usually that of the direction of tape motion. Both DC and AC fields of 1000 to 5000 Oe are usually applied; the agitation caused by the AC field is apparently helpful to maintain a uniform particle dispersion. The magnitude of the field and the time of the field application during the drying process are carefully controlled important parameters which determine the final performance of the coating.

The processes described up to now were carried out on a wide web of base material. The tape is eventually slit to carefully controlled final dimensions, wound onto the reel upon which it will be used and inspected and tested. Computer-grade tapes are often 100% tested for defects, in particular for 'dropouts' and noise pulses, often termed as 'drop-ins', recording current uniformity and output level. Other fundamental parameters – mechanical, magnetic – are usually tested on sample-per-batch basis.

2.10. TYPICAL TAPE PARAMETERS

The requirements for interchangeability of tapes not only between machines of one manufacturer, but between any two transports which conform to the same standard on recording format necessitate maintaining close tolerances on critical mechanical and magnetic properties of the recording medium. Figure 2.2. gives a list of the most important parameters for computer-grade tape.

Most of the parameters are self-explanatory, but as various national and international standards define slightly different test conditions, or do not strictly specify the method of measurement, the published figures in the manufacturer's data sheets need careful interpretation. Coercivity is determined from the hysteresis loop of the tape taking it up to 1000 Oe at 50 Hz. Output signal levels are often specified relative to a 'standard tape', and skew as the time lag between the two edge-track read pulses relating to the same character at 112.5 in tape speed (caused by the tape only and not by the imperfection of the recording system). It is necessary to avoid completely filling the reel with tape to prevent spillage. The margin between the outer convolution of the tape and the outside edge of the reel – which depends on the amount of tape on the reel and overall tape thickness – is specified by the manufacturer (for nominal tape length) as the 'E value'.

Tape parameter	Typical value
Coercivity	230 to 300 Oersted
Residual saturation flux	1.25 to 1.45 Maxwell
Permanent dropout	Nil
Drop-in	Nil
Print through	<1%
Base material thickness	1 to 1.5 mil (1000 to 1500 μin)
Coating thickness	0.4 to 0.6 mil (400 to 600 μin)
Total tape thickness	1.5 to 2 mil (1500 to 200 μin)
Tape width ($\frac{1}{2}$ in nominal)	0.498 ± 0.002
E value	$\frac{1}{4}$ to $\frac{1}{8}$ in
Cupping	< 0.1 in
Yield force	10 lb minimum
Coating roughness	1 to 6 μin
Surface resistance per square inch	0.5 to 100 MΩ
Linear coefficient of thermal expansion, 10°C to 50°C, 20% to 80% RH	35 x 10^{-6} per °C
Linear coefficient of humidity expansion, 10° to 50°C, 20% to 80% RH	11 x 10^{-6} per % RH
Frictional drag, coating on brass	130 g max*
Dynamic skew	< 2 μs*
Length tolerance, 2400 ft.	+50 ft, −0
Ambient conditions, operating	15.5 − 32.2°C (60 − 90°F) 20% to 80% RH
Ambient conditions, storage	4.4 − 32.2°C (40 − 90°F) 20% to 80% RH

* Test procedure as in ANSI x 3.2. 1/402.

Fig. 2.2. Typical computer tape parameters.

'Cupping' defines the transverse curvature (perpendicular to the longitudinal direction) to the tape* (Fig. 2.3). Permanent transverse curvature can create uneven head wear, poor uniformity of output between tracks. The tests are usually conducted on a $\frac{1}{4}$ in length of tape which is laid on a precision surface plate and peak deviation from the baseline is measured. Alternatively the radius

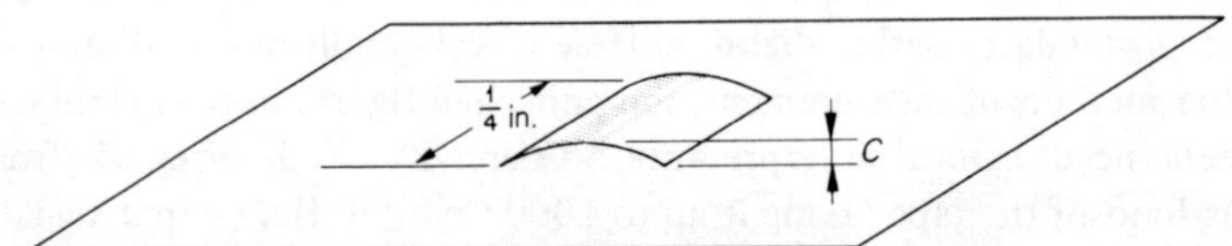

Fig. 2.3. Test of transverse curvature (cupping).

of the circle is specified of which the cross-section of tape is an arc. Cupping may be caused by incorrect drying or curing of the coating during the manufacturing procedure or because of the differences between the coefficients of thermal or hygroscopic expansion of the coating and base material.

* Concept is synonymous with the camber of a road surface.

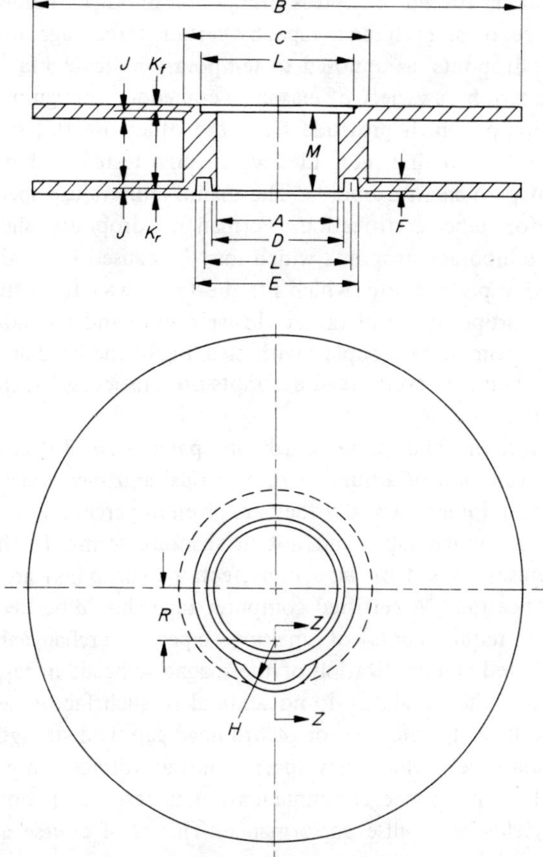

Fig. 2.4 (a). Tape reel.

Dimension	Inches
A	3.688
B	10.500
C	5.125
D	3.875
E	4.388
F	0.250
H	0.750
J	0.102
K_f	0.125
K_r	0.080
L	4.125
M	0.718
R	1.677

Fig. 2.4b. Reel dimensions.

The term 'dropout' is used to express a reduction in output below the average for the roll of tape, or of a reference tape, by certain percentage, usually 50% or 35%. 'Permanent dropouts' as opposed to temporary indicate a fault in the tape which may be caused by a variety of reasons – embedded foreign particles in the coating, oxide clumps which protrude from the otherwise flat surface, minor discontinuities in the coating, etc. Ideally, a fully tested and certified tape should be free of permanent dropouts. The manufacturers use special 'dropout test' machines for tape certification. Permanent dropouts should be distinguished from temporary dropouts which may be caused by a minute particle of dust, loose oxide particle, etc. which lift the tape away from the read head. These temporary dropouts are of rather elusive nature and re-reading the same area of the tape – sometimes coupled with cleaning of the head and tape guides – may eliminate them. If several read attempts are unsuccessful, the error is of 'permanent' nature.

Output signal levels and pulse width are parameters difficult to specify because they are function of a number of electrical and mechanical parameters of the write/read equipment. Usually they are given in percentage of the relevant parameter of a 'reference tape' and not in absolute terms. In the same way random noise pulses caused by tape imperfections (drop-ins) are specified in terms of a reference tape. A certified computer tape should be free of drop-ins.

Although the requirements of making tapes interchangeable between machines necessitated standardization of the magnetic heads in respect of track width and position, the standards do not extend to such factors as gap length, number of turns of read/write coil, or record head gap field strength. Hence the magnetic tape data sheets which may specify output voltage and pulse width of the replay signal do not make recommendation in respect of 'optimum' write current, which yields best pulse performance. This is of course a function of record head parameters, head-to-tape contact.

Figure 2.4 shows the main dimensions of a typical reel used for storing computer tape. The moment of inertia for an empty 10.5 in diameter reel is in the order of 0.02 in/lb/s^2.

2.11. CARE AND HANDLING OF MAGNETIC TAPE

Care of magnetic tapes is a major factor in obtaining optimum utilization and maximum life span. It includes a number of factors from operator education and preparation of the working area to correct adjustment of the tape transport.

The location in which the tape is used should approach the 'clean room' environment, kept dust-free and have a reasonably controlled temperature and relative humidity. Even a short period of operation in a dusty atmosphere can cause, if not irreparable damage, a large number of errors. Small particles on the tape surface disturb the head-to-tape contact and cause elusive temporary dropouts. One of the frequent mistakes in computer-room layout is to place paper tape equipment – punched card and punched paper tape – and magnetic tapes in the vicinity of each other exposing the magnetic tape equipment to the dust generated by the paper handling machinery. The transport manufacturers provide various tape-cleaning devices such as a miniature 'vacuum-cleaners' in the tape path and a separate, often pressurized compartments for the tape-handling area; this however does not obviate the necessity of near-clinical cleanliness in handling of tape. A typical clean-room dust specification calls for 0.1 to 1 mgm per cubic meter and 3 to 5 μm maximum particle size; ambient temperature shall be between 16°C and 32°C with 40% to 60% relative humidity. Large or rapid change of temperature should be avoided. The oxide surface of the tape should not be touched by hand; some machines of more recent design provide cartridge loading facilities where the tape loading is fully automatic.

2.12. TAPE LIFE

In well-designed transports it is only the magnetic head and sometimes a tape cleaner which is in contact with oxide-coated side of the tape. Nevertheless, after a certain number of passes the tape starts to deteriorate, not so much because of the loss of coating, but more because of minor particles tend to be embedded into the recording surface. The head itself consists of layers of materials with different hardness and abrasive properties, and iron oxide is a hard and abrasive material. The wear products, consisting of microscopic pieces of coating, head material, etc. tend to build up slowly on the heads and guide rollers and may get re-attached to the tape surface. The tape itself can be re-conditioned on special purpose tape cleaner machines which mechanically clean the recording surface and remove loose particles as long as the particles are not too firmly attached to the surface. The sign of tape deterioration is the increase in numbers of errors caused by dropouts. Practically every material and process utilized during the manufacturing process, the base material, binder, surface coefficient of friction, etc., does effect the normal life and this is even more so for operating conditions, tape handling, tension, temperature, humidity, tape path on the machines, and dust content of ambient. For near-optimum operating conditions as many as 10 000 passes over the head could be expected before significant increase in dropouts; in a dusty atmosphere or poorly adjusted tape transport the tape may deteriorate after a few hundred passes.

An incorrectly adjusted transport can damage the tape in a number of ways. If the tape path is incorrectly aligned so that the tape gets unevenly stretched, the edge of the tape becomes 'wavy'. This causes the outer tracks to loose contact with the head and consequently an increasing number of dropouts.

If tape tension is too low, rapid acceleration or deceleration – particularly at the end of a rewind phase – may cause the tape to slip with respect to the hub to which the braking force is applied. This spillage may cause a permanent crease in the tape with the consequence of dropouts. In poorly designed tape transports a loss of mains voltage may cause the sudden application of mechanical brakes with the consequence of stretching the tape or uneven winding. A protruding layer can get damaged very easily.

Figure 2.5 shows the general performance of a carefully handled reel of magnetic tape over its life. During the first few runs there may be some temporary errors as the result of surface irregularities which are smoothed out by the friction of the tape over the head and guides. The error rate remains not at zero but at very low level for many thousands of passes after which,

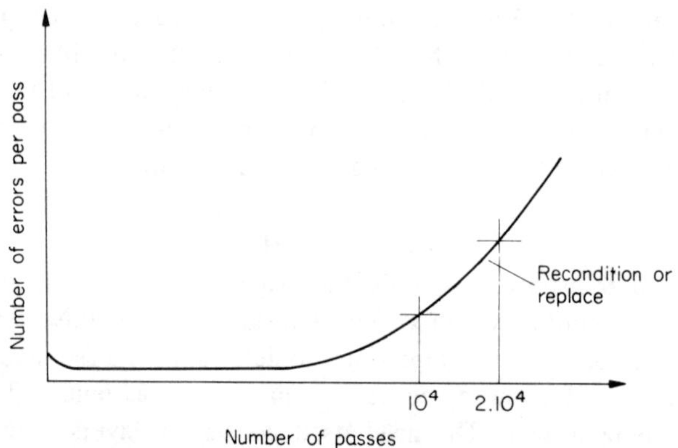

Fig. 2.5. Tape performance versus number of passes.

depending on the tape quality and the treatment the tape is subjected to by the tape handler, the error rate shows an increasing tendency. At a certain error rate the data throughput slows down because the error correction logic which is built into either the tape handler or the computer will try to re-read the block in which the error was detected. Eventually the tape must be withdrawn from service. Worn tape can be re-conditioned on purpose-built conditioning machines which remove loose oxide and small dirt particles from the recording surface.

REFERENCES

1. Poulsen, V. (1900), 'The Telegraphone', *Ann. d. Phys.*, **3**, 754–760.
2. Schüller, E. and Meyer, E. (1932), 'Magnetic sound recording on steel tape', *Z. f. Tech. Phys.*, **13**, 593–599.
3. Speliotis, D. E., Morrison, J. R. and Judge, J. S. (1965), 'A correlation between magnetic properties and recording behaviour in metallic chemically deposited surfaces', *IEEE Trans. Mag.*, **Mag-1**, 348–352.
4. Speliotis, D. E., Morrison, J. R. and Judge, J. S. (1966), 'Correlation between magnetic and recording properties in thin surfaces', *IEEE Trans. Mag.*, **Mag-2**, 3, 208–212.
5. Davies, A. V., Middleton, B. K. and Tickle, A. C. (1965), 'Digital recording properties of evaporated cobalt films', *IEEE Trans. Mag.*, **Mag-1**, 4, 344–348.

6. Bonyhard, P. I., Davies, A. V. and Middleton, B. K. (1966), 'A theory of digital magnetic recording on metallic films', *IEEE Trans. Mag.*, **Mag-2**, 1, 1–5.

7. Middleton, B. K. (1966), 'The dependence of recording characteristics of thin metal tapes on their magnetic properties and on the replay head', *IEEE Trans. Mag.*, **Mag-2**, 3, 225–229.

8. Nishikawa, M. (1968), 'Digital recording properties of relatively thick magnetic medium', *IEEE Trans. Mag.*, **Mag-4**, 286–290.

9. Morrison, J. R. (1968), 'A study of the effect of remanence and thickness on recording properties of thick particulate media', *IEEE Trans. Mag.*, **Mag-4**, 281–286.

10. Bate, G. (1965), 'High density recording on thin metallic tapes', 1965 Intermag. Conference, p. 12.1; *IEEE Trans. Mag.*, **Mag-1**, 193–205.

11. Bate, G., Morrison, J. R. and Speliotis, D. E. (1964), 'High density digital recording on thin metal surfaces', *International Conference on Magnetic Recording*, London, 7–8.

12. Speliotis, D. E. (1969), 'The effect of remanence in digital recording', *IEEE Trans. Mag.*, **Mag-5**, 3, 253–258.

13. Mee, C. D. (1964), *The physics of magnetic recording*, Chapters 5 and 6, North Holland Publishing Co., Amsterdam.

14. Mee, C. D. (1964), 'Magnetic tape recording materials', *IEEE Trans. Audio*, **Au-12**, 4, 72–82.

15. Speliotis, D. E. (1967), 'Magnetic recording materials', *J. Appl. Phys.*, **38**, 3, 1207–1214.

16. Mallinson, J. C. (1971), *Magnetic properties of materials*, Jan Smit (editor), McGraw-Hill.

17. Bate, G. and Alstad, J. K. (1969), 'A critical review of magnetic recording materials', *IEEE Trans. Mag.*, **Mag-5**, 4, 821–839.

18. Eldridge, D. F. (1965), 'Magnetic tape production and coating techniques', *Memorex Monograph*, **4**, 1965.

19. Daniel, E. D. and Eldridge, D. F. (1967), *Magnetic recording for science and industry*, C. B. Pear, Jun. (editor), Reinhold.

20. Morrish, A. (1965), *Physical principles of magnetism*, Chapter 7, Wiley.

21. Pearce, R. R. (1964), 'The measurement of the magnetic properties of recording tape', *International Conference on Magnetic Recording*, London, 18–23.

22. Mee, C. D. (1958), 'Magnetic tape for data recording', *Proc. Inst. Electr. Engrs. (BG)* **105**, **Pt. B**, 373–382.

23. Helbren, N. J. (1971), *The abrasivity of magnetic recording tape*, Fulmer Research Institute, Stoke Poges.

24. Davidson, P. S., Giles, P. and Matthews, D. A. R. (1968), 'Ageing of magnetic tape', *Computer Journal (GB)*, **11**, 3, 241–246.

25. Aharoni, A. and Fisher, R. D. (1965), 'Dependence of packing density on coercivity and thickness of recording tapes', *Proc. 1965 Intermag Conference*, Washington, D. C. 12.3-1–12.3-5.

26. Speliotis, D. E. (1968), 'A digital recording study of CrO_2 particulate media', *IEEE Trans. Mag.*, **Mag-4**, 553–557.

27. Speliotis, D. E. and Kump, H. J. (1964), 'Fundamental criterion for recording on magnetic surfaces', *Proc. 1964 Intermag. Conference*, 3-2.1-3-2.4.

28. Morrison, J. R. and Speliotis, D. E. (1965), 'The optimum write current for digital recordings on particulate oxide and metallic recording surfaces', *Proc. 1965 Intermag. Conference*, 12-4.1-12-4.8.

29. Wohlfarth, E. P. (1964), 'A review of the problem of fine-particle interactions with special reference to magnetic recording', *J. Appl. Phys.*, **35**, 783–790.

30. Ardenne, M. V., Effenberger, M., Müller, M. and Völtz, H. (1966), 'Investigations of the preparation and properties of evaporated magnetic films tape storage media', *IEEE Trans. Mag.*, **Mag-2**, 3, 202–205.

31. Krones, F. (1960), 'Preparation and elektro-acustic properties of magnetic tape stores for audio recording', *Technik der Magnetspeicher*, F. Winkel (editor), Springer, Berlin.

32. Famulener, K. (1967), *Encyclopaedia of chemical technology*, Kirk-Othner (editor), Wiley, New-York, **12**, Chapter 7.

33. Woodward, J. G. and DellaTorre, E. (1960), 'Particle interaction in magnetic recording tapes', *J. Appl. Phys.*, **31**, 1, 56–62.

34. Cummings, R. A. and Mee, C. D. (1964), 'The effects of particle orientation on interactions in magnetic recording tapes', *Proc. Intermag. Conference IEEE, Washington, D.C.*

35. Daniel, E. and Levine, I. (1960), 'Experimental and theoretical investigation on the magnetic properties of iron oxide tape', *J. Acoustic Soc. America*, **32**, 1, 1–15.

36. Camras, M. (1948), 'Magnetic recording tape', *AIEE Transactions for 1948*, Paper 48–75.

37. Carrol, J. F. Jun. and Gotham, R. C. (1966), 'The measurement of abrasiveness of magnetic tape', *IEEE Trans. Mag.*, **Mag-2**, 1, 6–13.

38. Camras, M. (1954), U.S. Patent 2694656.

39. Daniel, E. D. and Levine, I. (1960), 'Determination of the recording performance of a tape from its magnetic properties', *J. Acoust. Soc. Amer.*, **32**, 2, 258–267.

40. Woodward, J. G. and DellaTorre, E. (1959), 'Magnetic characteristics of recording tapes and the mechanism of the recording process', *J. Audio Eng. Soc.*, **7**, 4, 189–196.

41. Moradzeh, Y. (1966), 'Chemically deposited cobalt-phosphorus films for magnetic recording', *J. Electrochem. Soc.*, **112**, 9, 981–986.

42. Grimwood, W. K., Horak, J. R., Krall, H. J. and Delamore, P. J. (1967), 'Relationship between magnetic tape properties and magnetic oxide concentration', *IEEE Trans. Mag.*, **Mag-3**, 1, 49–53.

43. Siakkou, M. (1969). 'Influence of coating permeability in thin film pulse recording', *IEEE Trans. Mag.*, **Mag-5**, 4, 891–894.

44. Lissberger, P. H. and Comstock, R. L. (1971), 'Orientation of magnetic particles in a recording medium', *IEEE Trans. Mag.*, **Mag-7**, 2, 259–262.
45. Greiner, J. (1963), 'Anisotropy in magnetization of magnetic tapes', *Hochfrequenz u. Elektroakustik*, **72**, 2, 54–63.
46. Schmidt, E. and Franck, E. W. (1953), 'Manufacture of magnetic recording materials', *J. Soc. Motion Picture Television Engrs.*, **60**, 453–463.
47. Seidel, B. (1972), 'Chromium dioxide magnetic tape', *Fernseh & Kino-Tech.* (Germany), **26**, 7, 227–232.

3 The magnetic head

The 'writing' process in a digital magnetic recording system can be defined as the conversion of coded information represented by electrical pulses into pattern of magnetization on the recording medium. The current-to-magnetic field transducer – which has become known as magnetic head – is a device unique to magnetic recording. Limitation of the recording and reproduction process are directly related and critically dependent on the characteristics of the magnetic head.

Some familiarity with the construction of magnetic heads is necessary before we can embark on the theory of recording. In this chapter we will review the materials, technology, construction and design of multi-track tape heads.

The thin layer of magnetizable material carried on a plastic base can be magnetized in three principle directions: longitudinally along the tape, transversly in the direction of the width of the tape, and perpendicularly to the thickness of the layer. The recorded magnetization generated in the tape for each of the three principal directions is schematically illustrated in Fig. 3.1. Longitudinal magnetization permits a better utilization of the recording surface than the others and, with the exception of video recording, all applications – audio, instrumentation, digital – use longitudinal recording.

Functionally the heads fall into three categories: write, read and erase heads. The basic construction of all the three types is very similar. The conventional magnetic head – and continuous efforts over three decades have not yet replaced this type – consists of a ring-like structure with a magnetically soft high-permeability core and a winding (Fig. 3.2).

For convenience of manufacturing and coil winding the ring-like structure is usually made of two halves; consequently the structure has not only a front gap – at which the writing and reading takes place – but a rear gap. The rear gap appears as an undesirable increase in the core reluctance hence it is kept as short as possible. The dimensions of the front gap are very finely controlled because the length and depth of this gap define (other factors being equal) the overall system performance. An enlarged view of the pole tip area of the head with an

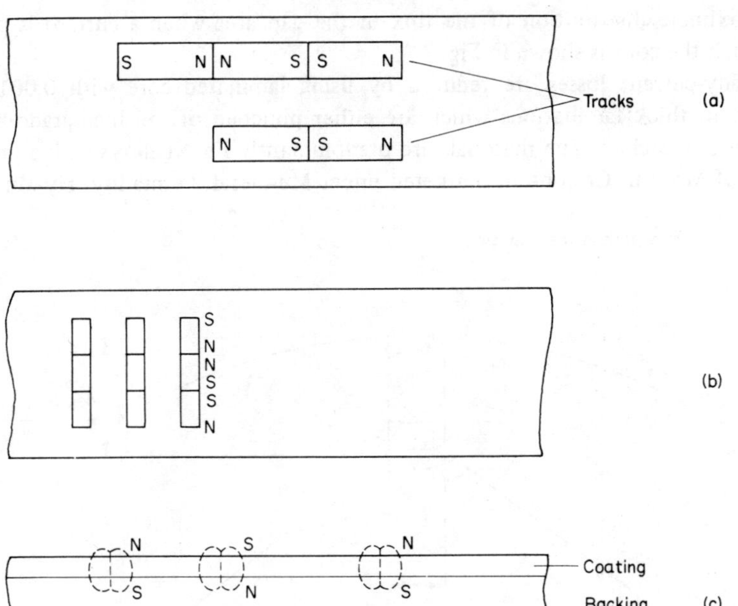

Fig. 3.1. Longitudinal, transversal, perpendicular recording;
(c) shows cross-section of the recording medium.

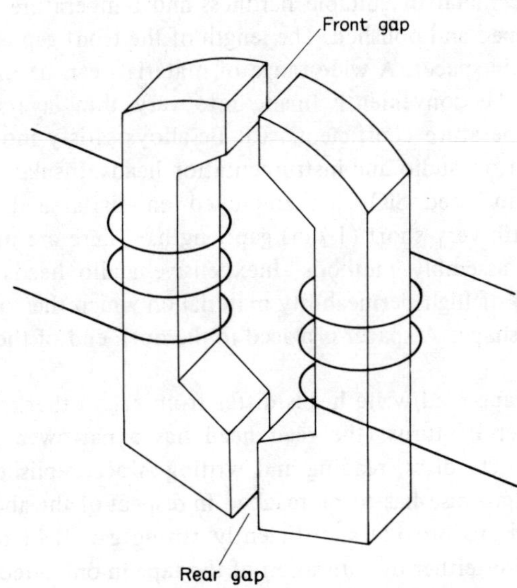

Fig. 3.2. Schematic of a write/read head.

approximate distribution of the flux in the gap area when a current is passed through the coils is shown in Fig. 3.3.

Eddy-current losses are reduced by using laminated core with 0.001 in – 0.005 in thick laminations which are either punched, or, on high-grade heads, chemically etched. The materials are predominantly Fe–Ni alloys with a few per cent of Mo, Cu, Cr content marketed under Mu-metal, Permalloy, Hy-Mu trade

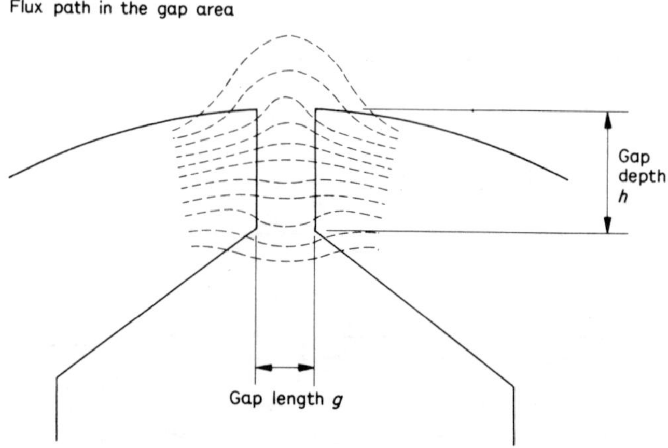

Fig. 3.3. Enlarged view of the pole-tip area of a write head.

names. The annealed laminations are bonded into stacks after which the coils are wound on to them. This sub-assembly is then fitted into a housing which is made of a non-magnetic metal of suitable hardness and temperature coefficient; the two halves are lapped and polished. The length of the front gap is determined by a non-ferromagnetic spacer. A wide range of materials can be used as spacer as long as they can be conveniently made into very thin layers, have suitable hardness and temperature coefficients. Cu–Be alloys satisfy most requirements and are often used on audio and instrumentation heads. Insulators, in particular mica, glass or evaporated SiO_2 are employed on digital and high-frequency analogue heads with very short (1 μm) gap lengths. There are many alternative construction and assembly methods. Inexpensive audio heads are fabricated from a single strip of high permeability material on which the coil is wound and then bent into U shape. A spacer is placed in the open end of the U forming the read/write gap.

The magnetic tape read/write heads differ from each other in respect of gap length and number of turns (the read head has a narrower gap and higher number of turns). On discs, reading and writing is accomplished by the same head hence a compromise has to be reached in respect of the above parameters. The erase head has to provide a sufficiently strong gap field to obliterate all previous information either by saturation of the tape in one direction (DC erase) as in digital recording or leaving the tape in a magnetically neutral status which is obtained by exposing the tape to a sufficiently strong and gradually reducing AC field (analogue recording). The erase head is again of the same construction for either AC or DC erasure as the read/write head but may employ different core materials (high saturation intensity of magnetization is required) and a wider gap.

3.2. MULTI-TRACK HEADS

The construction of a single-track, audio read-write head is relatively simple and permits the use of mass-production techniques. However when multi-track heads have to be built to the very close tolerances which are required for high-density digital and high frequency analogue recording, head design and production becomes a challenging problem of precision electro-mechanical engineering. This is reflected by the immense number of patents covering all aspects of design, construction, manufacture and testing of precision magnetic heads.

Multi-track heads may be built up from single-track heads track by track with the gaps optically aligned at the final assembly stage; alternatively two halves of a stack are built up separately and eventually brought together. In either case the crosstalk – interaction of the closely spaced tracks – has to be prevented by appropriate choice of the winding sense on adjacent coils and in particular by an inter-track screen. The shape of the screen and the choice of the screening material – particularly in high-frequency analogue recording – needs careful consideration. The inter-track screen becomes a part of the electromagnetic structure of the head and influences the head impedance and high-frequency response [1, 2, 3, 4, 5, 6]. Often a 'sandwich' of ferromagnetic and good electrical conductor is chosen such as copper-plated Fe–Ni alloy.

High performance digital tape systems use a dual-gap read-after-write head for checking the information just written. The head stacks in the dual gap head are formed of C and I or L sections and the two heads are separated by a narrow centre section (Fig. 3.4). It would be desirable to place the two gaps so close to each other that the bit just written could be checked by the read head before writing the next digit. Such systems were implemented at low recording densities with a head construction which permits the read/write gaps to be very close to

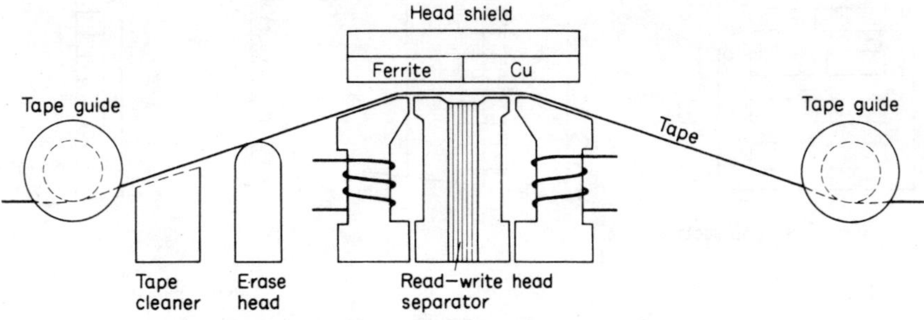

Fig. 3.4. Schematic of a dual gap write/read head assembly.

each other [8, 9, 10] but are not practical at higher densities. The conventional head construction necessitates a minimum distance of about 0.15 in between the read and write gaps in a dual gap head.

A special problem associated with the dual-gap heads is the elimination of 'crossfeed', i.e. the direct coupling of the high-level recording pulse in the adjacent read head coils [60, 61]. The read and write sections of the dual gap head are separated by a laminated screen; the position and sense of windings is chosen to minimize coupling. A further crossfeed-reducing measure is the use of a shield in front of the face of the head (Fig. 3.4). This shield is often constructed from a combination of a non-ferromagnetic good electrical conductor and a ferromagnetic material such as copper and ferrite. The distance between the shield and the face of the head is sufficient to permit free passage of the tape; the shield has a hinged mounting permitting its removal during tape loading. The construction of a multi-track head must permit to maintain very close tolerances on mechanical dimensions and electrical parameters at a realistic cost. The maximum recording density which can be obtained with an acceptable error rate in a recording system is critically dependent on the head parameters such as read head gap length, gap scatter, deviation of the gap centre line from its nominally perpendicular position to the base plate in the direction of the tape motion (azimuth angle) the tilt of the head stack perpendicular to the direction of the tape motion, uniformity of track width, and track position. The terms gap scatter, azimuth, and tilt are illustrated in Fig. 3.5. The coil impedances must be carefully matched otherwise the variation in write current rise time and the corresponding scatter in the position in the recorded bit limits the maximum bit density.

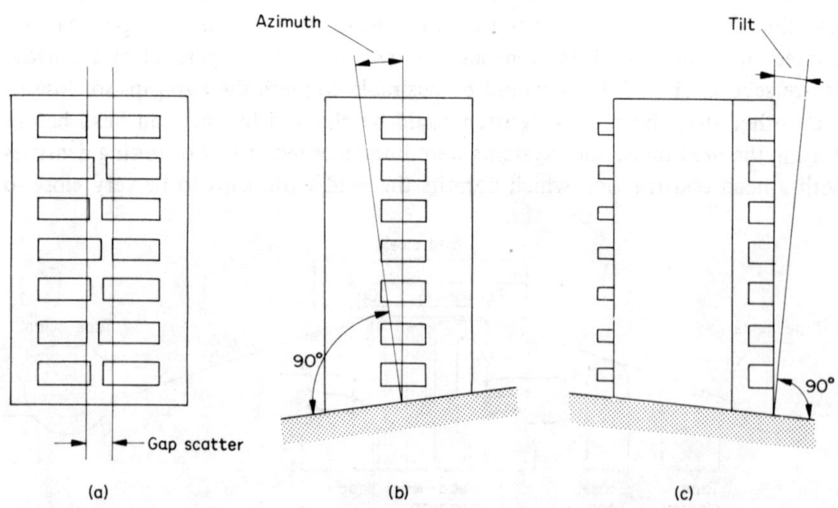

Fig. 3.5. Gap scatter (a), azimuth (b), tilt (c).

The gap scatter is defined as a band around the ideal gap centre line, containing both edges of all gaps in one headstack. Ideally all gaps should be perfectly in line; in practice the gap scatter is in the order of 20 to 100 μ in. Various standards specify a maximum permissible value. Although gap scatter can be compensated electronically it is one of the practical limiting factors of the recording density.

The considerations which dictate the choice of the read and write head gap length will be discussed in Chapter 4. The main dimensions of a typical dual-gap read/write head are shown in Fig. 3.6. For saturation digital recording write gap length is usually less than 500 μ in, the read head between 200 and 75 μ in. High-frequency instrumentation, video or even audio heads may have shorter gaps and head with 40 μ in (1μm) read gap length are in use.

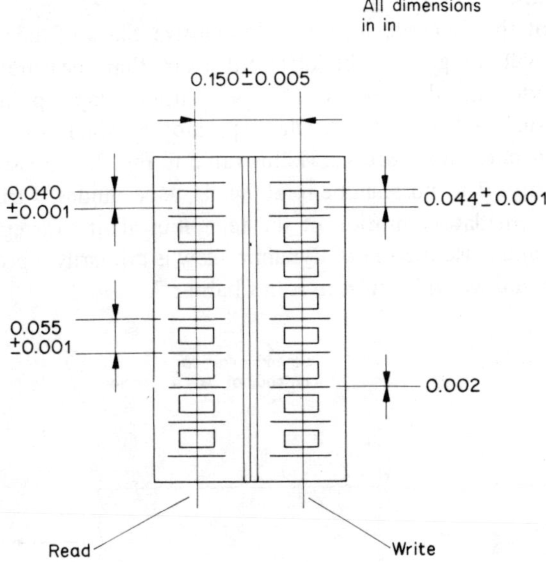

Fig. 3.6. Nine-track dual-gap read/write head. Typical dimensions and tolerances.

3.3. SKEW

The imperfections of the write/read heads, tape and transport cause a deviation of the simultaneously recorded bits in one character (a group of 6 or 8 bits) from time coincidence during reading. As this skew* is a practical bit density limiting factor, keeping it at the lowest possible value is of paramount importance.

* Skew: 'The deviation of bits within a tape character from the intended or ideal placement which is perpendicular to the reference edge'. Definition in ANSI X3.2 1/400.

The skew can be divided into a static component that is unvarying and is dependent on the constant parameters of heads, transport and tape and a dynamic component which varies from character-to-character. Clearly the constant, static component can be largely compensated by relatively simple, mechanical or electronic means; reduction of dynamic skew is the more complex problem.

The dominant component of the static skew originates from the misalignment of the gaps in a multi-track head; compensation can be achieved by addition of a variable electronic delay circuit to each track.

The tape characteristics which may contribute to the skew are minute variation in tape width (slitting defects), cupping tendency, non-uniformity of tape thickness, variation of surface friction. On the transport the main causes are minute irregularities in the tape path caused by imperfectly aligned tape guides, capstan and heads.

The nature of the dynamic skew is rather elusive. Skew of tape, measured on one transport will, in general, be different from that measured on another transport because the differences in tape tension, tape path, tape guide tolerances and surface frictions. Irregular tape motion which is the consequence of tape and transport imperfections is illustrated in Fig. 3.7, grossly exaggerated for sake of clarity. The misalignment of heads, tape guides, variation in tape width causes an oscillatory motion of the tape equivalent to irregular variation of the azimuth angle. Reduction of dynamic skew is primarily a problem of tape transport design and we will return to it in Chapter 7.

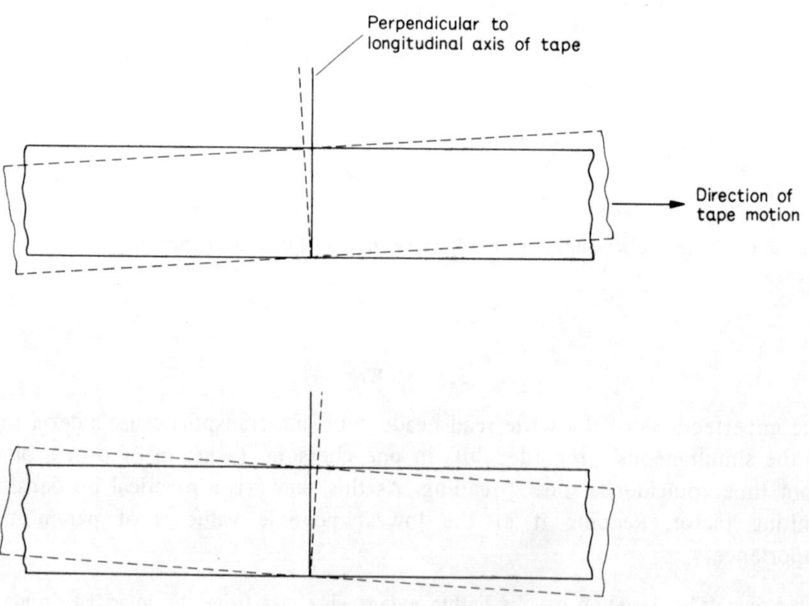

Fig. 3.7. Tape motion irregularity.

3.4. ERASE HEADS

Digital tape recorders usually employ a full-width erase even in saturation recording where the most recent information obliterates the previous one. If a full-width erase head is not used and a recorded tape is transferred to another machine, the variation in tape guidance between transports may permit the previously recorded information to remain present beyond the edge of the track during over-writing with new data. Subsequent reading can now pick up residuals of previously written information which appear as 'noise' on the latest data and can cause errors. Use of a full-width erase head eliminates this problem. The other reason for using a separate erase head is connected with the interblock gap. Previously recorded information must not be left in the interblock gap area. With the DC erase head and the write head producing saturation flux in the same direction the discontinuities at the edges of the interblock gap will be very small.

The mechanical construction of erase heads is much simpler than read/write heads as they are usually as wide as the tape and are not divided into tracks. Also, DC erase heads do not need to be laminated and ferrite head material is often used. The erasure field should be in the order of 1000 Oe for conventional oxide tapes throughout the full thickness of the medium; this is achieved by choosing a gap length in excess of the write head gap length and provision of sufficient current drive.

3.5. HEAD MATERIALS

The desirable magnetic and mechanical properties of the magnetic transducer head material are fairly well defined. First of all only magnetically soft materials which do not retain any permanent magnetization can be considered; any permanent magnetization could erase or weaken previously written information during re-reading. The write head material shall have high saturation intensity of magnetization to avoid saturation of the pole tip area – where the cross-section of the core is the smallest – by the write current. High electrical resistivity is desirable to keep eddy-current losses low. The permeability shall be sufficiently high and fairly constant throughout the frequency range of operation so that the core reluctance is low with respect to the air gap reluctance. The head material shall have good wear properties for in-contact recording on oxide surfaces as the oxides are hard and abrasive [2, 3, 6, 11, 12]. Finally there are such considerations as ease of manufacture and handling of thin laminations, variation in permeability after annealing caused by cold working, cost of the material and production cycle.

There are, for practical purposes, only two main categories of materials in use: Fe–Ni alloys and ferrites. Others which offered great promise because they combine good magnetic properties with mechanical hardness such as Alfenol, Sendust, have not gained general acceptance [13, 14].

The high-nickel alloys (Permalloy, Mumetal, Hy-Mu) have the highest permeabilities but are good electrical conductors – hence the need for laminated construction – and are mechanically soft [15, 16, 17].

Typical parameters are listed in Fig. 3.8. In fully annealed state the permeability can be in the order of 50 000 to 100 000. Any cold working, bonding, lapping can drastically reduce this high permeability so that great care is taken during the manufacturing process to keep cold working at minimum. Lapping, polishing the front gap area during manufacture – a process required in many designs to provide correct dimensions and surface quality – destroys the permeability in a thin surface layer with the consequence that the 'effective gap length' is slightly more than the gap spacer material. It is this effective gap length of the reproduce head which basically determines the high frequency response of the magnetic tape system.

Trade name	Composition, * % by weight	Relative permeability		Coercive force, H_c(Oe)	Saturation induction, B_s(G)
		Initial	Max		
Mumetal	Ni 77, Cu 5, Cr 2	30 000	70 000	0.03	8 000
Permalloy 4-79 (HY-MU 79)	Ni, 79, Mo 4	20 000	60 000	0.035	9 000
Supermalloy	Ni, 79, Mo 5, Mn 0.5	60 000	300 000	0.006	8 000
Ferroxcube	Fe_2O_3, ZnO, NiO	850	2 500	0.3–1.5	4 000
Alfenol 16	Al 16,	4 000	60 000	0.02	8 000

Actual values depend on frequency of measurement, lamination thickness or composition for ferrites, field strength for definition of initial permeability, hot or cold rolled, etc.

* Remainder: Iron + impurities.

Fig. 3.8. Properties of some typical magnetic head materials.

As the hard and abrasive oxide coating is continuously rubbing against the mechanically soft head material, the depth of the gap changes and eventually the heads 'wear out'. The severity of the problem depends on tape speed, the profile of the head, the head-to-tape pressure and the tape surface quality. The depth of the pole tip is dictated both by electrical and mechanical considerations. The record head pole tip must have a miminum cross-section area to avoid pole tip saturation by the record current (pole tip saturation would have the effect of drastic increase in gap length and incomplete saturation of the tape). At the other hand, the front gap reluctance must be large with respect to the core reluctance, hence the gap depth shall be small. Many different solutions have been tried, such as plating the pole tip area with a hard ferromagnetic material or making the pole tip exchangeable, but none of them has been fully successful. A novel solution [12] consists of hard ceramic 'wear bars' which provide support for the tape where it contacts and leaves the head, and allows the tape to follow its natural path across the active part of the head.

The application of ferrites in magnetic heads as an alternative to laminated metals was patented as early as 1950 [18]. Compared with Fe–Ni alloys, ferrites have lower saturation flux density, somewhat higher coercive force, much lower permeability and very much higher resistance (Fig. 3.8). The low permeability indicates that a ferrite core head, when used for writing, will need higher drive current than the corresponding Fe–Ni alloy head, and as a read head it will have lower efficiency. However, as some elementary calculations on the magnetic circuit of a typical digital head design will prove, the deterioration in performance caused by the lower permeability is not serious.

The high specific resistivity and corresponding low eddy-current losses make ferrite materials eminently suitable for very high frequency instrumentation and video applications, and for areas of out-of-contact recording where surface properties are less significant, i.e. drum and disc recorders. Manufacturing techniques differ from metal head processes and are described among others in detail by S. Duinker [19], H. P. Peloschek et al. [20], G. P. Bakos [21], E. Hirota et al. [22], H. Sugaya [24]. Good high-frequency response needs very sharp and well-defined gap edges; the area in contact with the tape must be highly polished. The difficulty in machining ferrite to the close tolerances required by multi-track heads has somewhat limited their application in tape, as distinct from drum and disc recorders, although ferrite heads are manufactured for audio applications and are used in all grades of domestic and professional audio and instrumentation recorders. Bonded glass spacers [19] can ensure a gap length as low as 1 μm; the ratio of effective to mechanical gap length is very close to unity.

Early ferrite heads suffered from erosion of the head surface where it was in contact with the tape. The mirror finish became pitted after a relatively short time. The phenomenon is explained by G. P. Bakos [21] by the fact that microscopic pores of the head surface were filled with some less firmly attached particles during the lapping operation. The loose particles attached themselves to the tape and caused tape contamination. Recent 'high density' ferrites seem to have solved this problem. Digital tape recorders rarely use ferrite read/write heads but often employ ferrite erase heads of full tape width.

3.6. DESIGN CONSIDERATIONS

The main design parameters of the write head can be derived in simple manner from the magnetic field distribution in the vicinity of the front gap. Making the simplifying assumptions that the magnetic field is constant in the z direction, the determination of field strength in the vicinity of the write head gap is reduced to a two-dimensional problem. A further idealization is to neglect the effect of the permeability of the recording medium on the head field which is equivalent to considering the medium permeability equal to the unity. The error will be small as the permeability of the conventional oxides is in the range of 1 to 3. Finally the write head core material is assumed to have a permeability approaching infinity. In this case lines of constant H (flux lines) will leave and enter the head surface at right angles (Fig. 3.9). The field at any point can be computed by taking the path of integration along a line of constant H from one pole to the other:

$$\oint H \, dl = Hl_a + H_{co}l_{co} = \frac{4\pi}{10} \, nI. \tag{3.1}$$

where

H = field strength along the path outside the core
l_a = length of path in air
H_{co} = field strength within the core assuming uniform cross-section.
l_{co} = path length in core
n = number of turns
I = write current

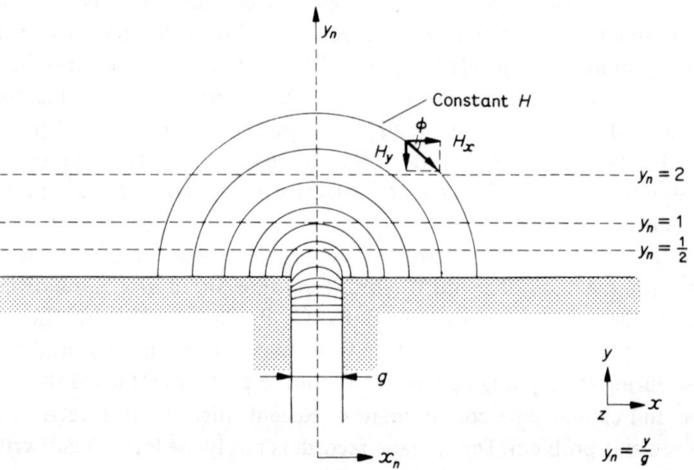

Fig. 3.9. The magnetic field in the vicinity of the gap of an ideal write head.

The permeability of the core material is sufficiently large to make $H_{co} = B/\mu_c$ negligible with respect to H so that Equation (3.1) reduces to

$$Hl_a \cong \frac{4\pi}{10} \, nI. \tag{3.2}$$

In the region deep inside the gap where the field distribution can be considered homogeneous the field strength will be:

$$H_{gap} = \frac{4\pi}{10} \frac{nI}{g}, \tag{3.3}$$

and for a path outside the gap

$$\frac{H}{H_{\text{gap}}} = \frac{g}{l_a}.$$ (3.4)

For the coordinates chosen in Fig. 3.9 $H_x = H \cos \phi$ and $H_y = -H \sin \phi$. The field plot is symmetrical about a plane through the centre line of the gap ($x = 0$) and the problem is treated as a static one with constant write current. Figures 3.10 through 3.13 show the field distribution in the vicinity of the gap of the

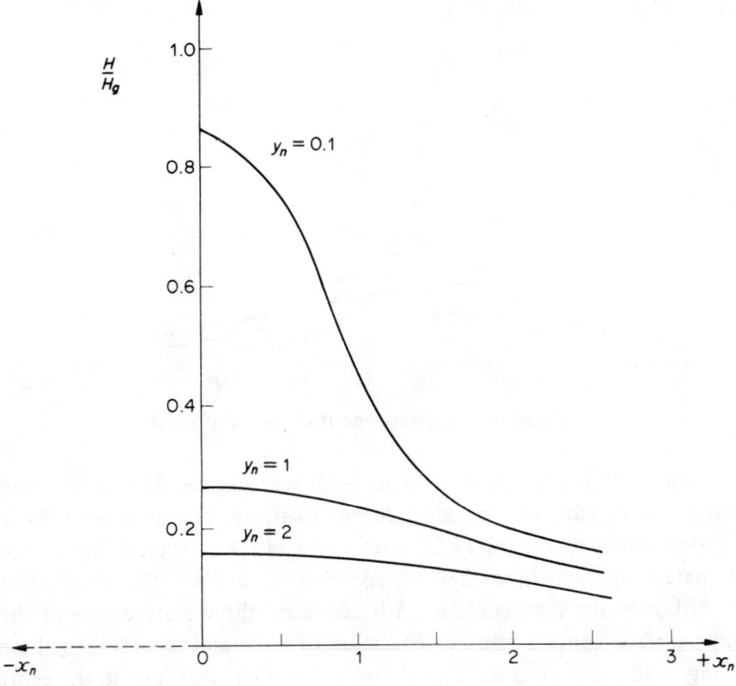

Fig. 3.10. Relative field strength in the vicinity of the gap of an ideal ring head.

idealized head against distances normalized with respect to the gap length $x_n = x/g$, $y_n = y/g$ [26, 27, 28]. It should be noted, that the y component of the field has opposite polarity on the two sides of the gap centre line whereas the x component does not change sign. It is the x component of the field which is of major importance for longitudinal recording. The recording medium is very thin therefore any elementary permanent magnet created in the medium by the y field of the head will be subjected to strong demagnetizing fields. (We will return to the problem of self-demagnetization in Chapter 4.)

Recording theory indicates that the gap length of the recording head has only a second-order effect on the recording process; if it is too narrow, only those layers of the medium will be saturated which are nearest to the head surface.

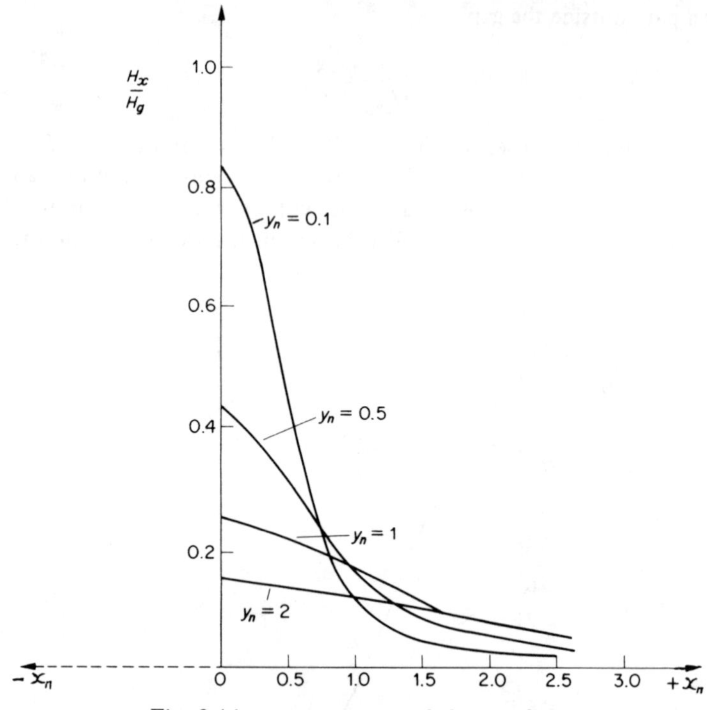

Fig. 3.11. x component of the gap field.

This is undesirable because of loss in readback voltage. If it is too wide the maximum bit density (bit packing density) suffers. Let us choose the record head gap length to be 0.5 mil ($1.27\ 10^{-3}$ cm) and take a typical digital recording oxide coated tape which has 280 Oe coercivity and 0.45 mil thick coating (Fig. 3.14). Although the tape is recorded 'in contact' the imperfections of the head and tape surface cause a minute separation of head and recording medium. The recording head must produce a field strength which is, at least in the centre line of the gap ($x = 0$) sufficient to saturate the full cross-section of the coating, i.e. the head field at $x = 0$ and $y = d + t_m$ distance must be greater than 280 Oe. The computed value of the x component of the field (Fig. 3.11) at $d + t_m = 0.5\ \text{mil} = g$ distance is approximately $0.28\ H_g$ thus the field deep in the gap must be $H_g = 280/0.28 = 1000$ Oe.

As a first-order correction the effect of the finite permeability can be taken into consideration:

$$H_{\text{gap}} = \frac{4\pi}{10} \frac{nI}{g} \frac{1}{1 + \dfrac{l_{co}A_g}{\mu_c g A_c}} \tag{3.5}$$

in which l_{co} is the magnetic flux path length in the core, μ_c the core permeability, A_g and A_c the gap and core cross-sectional areas, assuming

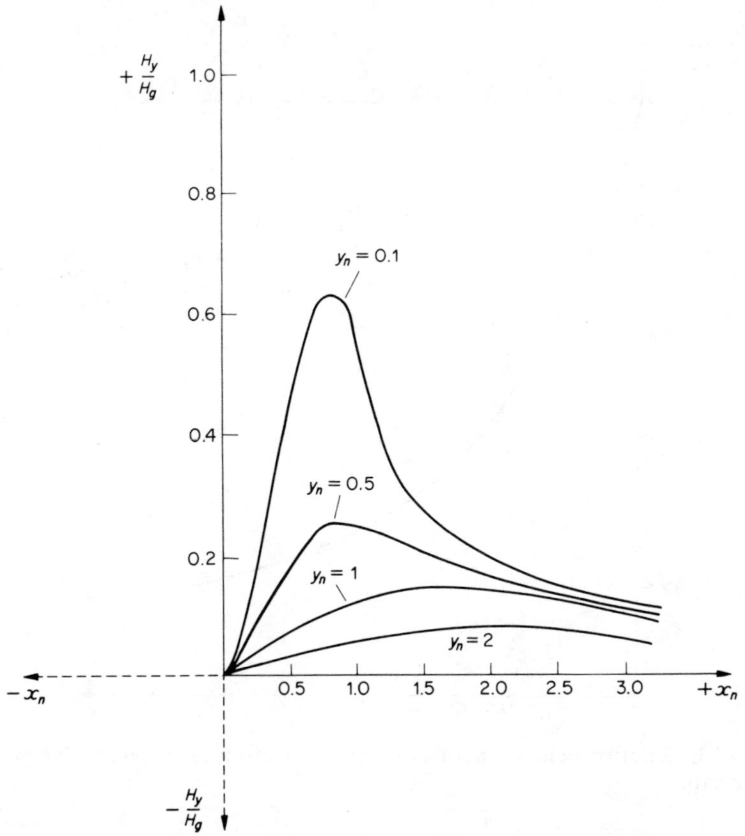

Fig. 3.12. y component of the gap field.

uniform core cross-section. The effect of the rear gap can be taken into consideration by a further correction factor. Once the gap field requirement has been established the magnetic circuit of the record head can be designed and the number of turns, write current, and head inductance calculated. In spite of the drastic simplifications used in derivation of Equation (3.5), it gives a good first approximation. Experience shows that for a typical digital recording tape, the optimum record current, defining 'optimum current' as the one which yields highest bit density in saturation recording with acceptable reproduce signal levels, is between 1 and 2 ampere-turns. A write head efficiency factor η_w can be defined as the ratio of magnetic field in the gap to the ampere-turns of the head:

$$\eta_W = \frac{H_g}{nI} \tag{3.6}$$

Using Equation (3.5) the write head efficiency can be computed as a function of permeability and geometry, remembering that the permeability is a function of

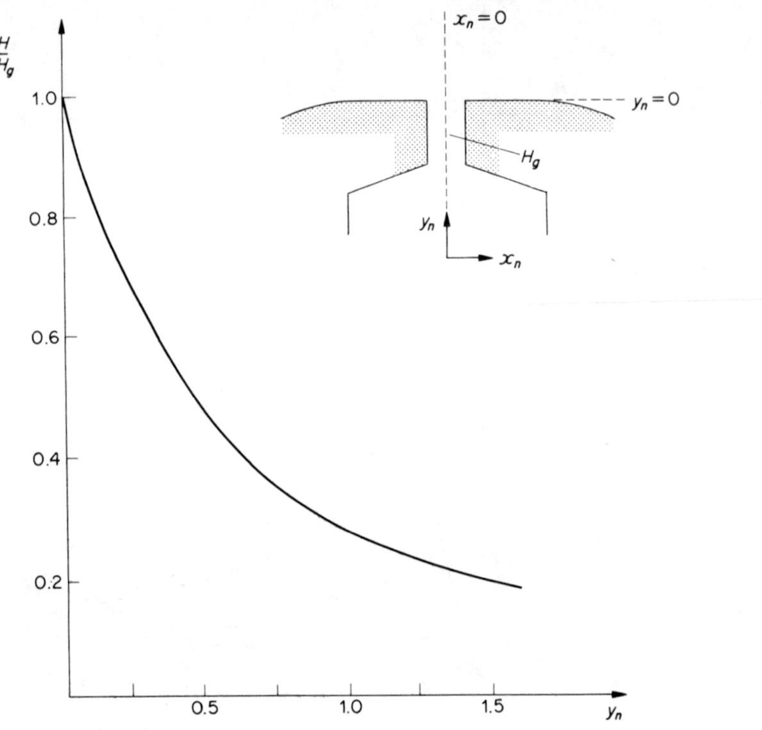

Fig. 3.13. Relative field strength as function of distance in centre line ($x_n = 0$) of the gap.

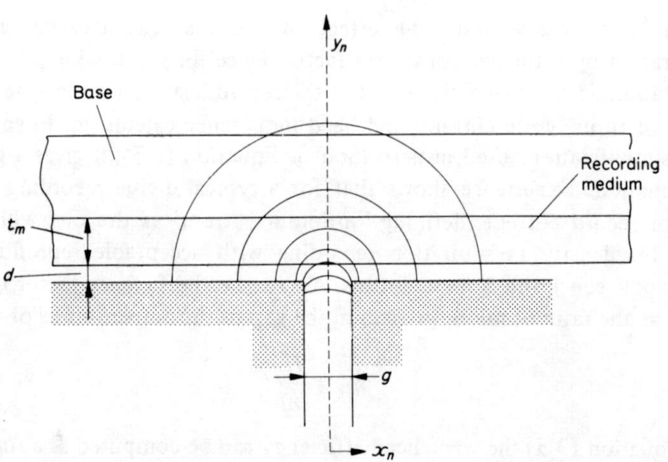

Fig. 3.14. Recording medium in the field of an ideal head.

the drive current. As the write head gap reluctance is large with respect to the core reluctance, the conventional write head designs are not sensitive to permeability variations.

The field plots in Figs. 3.10 through 3.13 give valuable information about some aspects of the recording process. As a particle is travelling along a line parallel to the x axis it will experience first a strong y field. Approaching the centre line the y field decreases and the x field increases; moving now past the centre line the y field changes polarity and starts increasing again while the x field remains of the same polarity. The recording field was required to saturate the medium at its farther boundary from the head ($d + t_m$ in Fig. 3.14). In this remote layer the change in the field strength in the x direction is slow; if the field strength is chosen to saturate the layer at the farther boundary in the centre line of the gap, the saturated area in this layer is restricted to the vicinity of the centre line. But at the same time the layers of the medium near to the gap ($y_n < \frac{1}{2}$) will experience a saturation field not only around the centre line but at a distance extending well beyond the gap edges in the x direction. Returning to our example in which $t_m + d$ was equal to g, the remotest layer of the medium is at $y_n = 1$ distance from the head surface. Saturation in the centre line at $y_n = 1$ extends the saturation region at $y_n = 0.1$ to about 1.8 gap lengths. Further, the layers near to the gap surface ($y_n < \frac{1}{2}$) will be 'over-saturated' in the vicinity of the centre line ($x_n = 0$). Clearly, it will be desirable to keep the thickness of the recording medium as small as possible to avoid the 'spread' in the recorded magnetization.

3.7. READ HEAD EFFICIENCY

The main read head design parameters are the length of the front gap, depth and shape of pole-tip area (Fig. 3.3), the choice of core material and cross-section, head inductance or number of turns, and track width. It is desirable to keep the rear gap length as low as possible although a non-negligible rear gap reluctance reduces the sensitivity of the design to material permeability variations. Once a few of the basic design parameters have been decided, others – inductance, output voltage for given magnetization, and head-to-medium speed – can be computed and the design can be optimized with respect to selected parameters. Finally, a compromise solution has to be found between desirable electrical parameters and producibility. A good insight can be gained on the effect of the permeability of the head material on head performance by defining a read efficiency factor η_R as the ratio of core flux to total flux.

The schematic illustration of the read process has shown that the flux emanating from the recorded tape closes partly through the low-reluctance path of the core, and partly through other routes. The induced voltage in the coil is proportional to the time derivative of the flux linking the core; this flux is the useful portion of the total flux, the rest is lost. The assumption can be made that the path for the 'lost' flux which does not couple into the head coil is through the front gap. The division of the total flux ϕ_t between gap and core will be inversely proportional to the reluctances of the two paths i.e. assuming uniform core cross-section we can write:

$$\phi_t = \phi_{gap} + \phi_{core}, \tag{3.7}$$

$$\frac{\phi_{core}}{\phi_{gap}} = \frac{R_{gap}}{R_{core}}. \tag{3.8}$$

Here R_g and R_c are the core and gap reluctances, respectively. Although care is taken to keep the rear gap small, it is not negligible and appears in series with the core reluctance:

$$R_{core} = \frac{l_{co}}{A_c\mu_c} + \frac{l_a}{A_a}. \tag{3.9}$$

Here l_{co} and l_a are the core and rear gap flux path lengths, A_c and A_a the corresponding cross-sectional areas and μ_c the (relative) permeability of the core material. The reluctance of the front gap is:

$$R_g = \frac{l_g}{A_g}, \tag{3.10}$$

where l_g is the front gap length and A_g the front gap cross-section area. From Equations (3.7) to (3.10) the read head efficiency is:

$$\eta_r = \frac{1}{1 + \dfrac{A_g l_{co}}{A_c \mu_c l_g} + \dfrac{A_g l_a}{A_a l_g}}. \tag{3.11}$$

Equation (3.11) indicates how the parameters of the magnetic circuit and head material permeability will influence the head efficiency. Figure 3.15 illustrates the effect of permeability variation on a typical read head; there is small deterioration in efficiency as the permeability is reduced from 50 000 to 5000 and little is gained by increase in permeability above 10 000. The design is insensitive to large changes in permeability because of the dominant effect of the rear gap reluctance in the core reluctance path [25, 29].

3.8. HEAD PROFILE AND HEAD-TO-TAPE SEPARATION

Magnetic tape systems as distinct from discs and drums nearly always use 'in-contact' recording aiming at zero head-to-tape separation. Other factors being equal, highest recorded bit densities, resolution and frequency response are obtained at the lowest possible distance between the heads and the medium. As the head-to-medium separation is a critical performance-determining factor it may be appropriate to investigate the influence of the head parameters on separation. Obviously, surface irregularities of tape and head, tape tension and

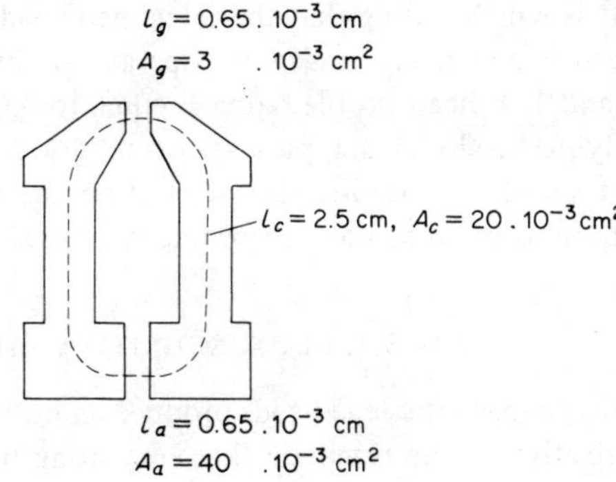

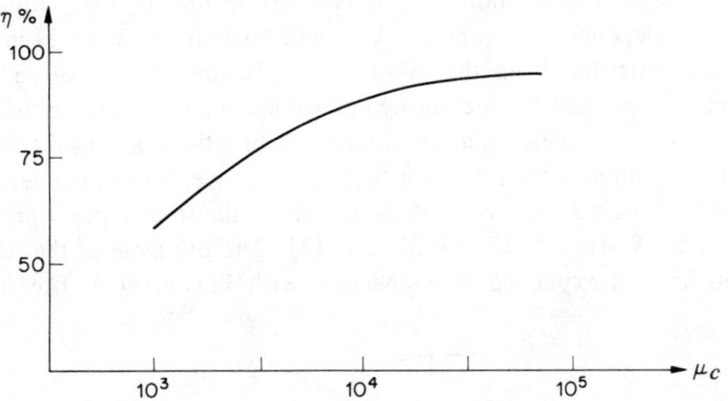

Fig. 3.15. Effect of permeability on the efficiency of a read head.

head profile will be factors of prime importance [70]. R. E. Norwood [30] analysed the complex interaction of various parameters taking into consideration the bending stiffness of the tape. As the tape is pulled across the head, a body of air will collect between head and tape so that the tape will actually float on an air cushion. The thickness of this air film as the function of tape tension, head surface radius and tape velocity was calculated – among others – by H. K. Baumeister et al. [31, 32] and J. A. Weidenhammer et al. [33]. A good approximation for the thickness of the air film is

$$h = 0.642 \, R \left(\frac{6\mu_a v}{T} \right)^{3/2}, \tag{3.12}$$

in which R is the head profile radius, v the velocity of the tape, μ_a the viscosity of air. The angle of tape wrap on each side of the tape is

$$\Theta = 16.8 \, \frac{h}{R} \left(\frac{T}{6\mu_a v} \right)^{1/3}. \tag{3.13}$$

Equation (3.12) predicts a linear relation between separation and head profile radius. It is worth noting that the calculated head-to-tape separation for 100 ips tape speed – and many transports operate at this or higher speed – 8 oz tape tension and 1 in head profile radius is equal to 85 μ in. This is the same distance as the flying height of a typical 'floating' disc head. The conclusion is that at high tape speeds the head-to-tape separation has to be taken into consideration even at nominally 'in-contact' recording systems.

3.9. FLUX SENSITIVE HEADS

Conventional magnetic read heads produce an output voltage proportional to the time derivative of the magnetic flux emanating from the tape, thus the output voltage – within certain limits – is proportional to the tape speed. Flux sensitive heads allow the reading of stationary or very slowly moving tape, and the head output is not dependent on speed. Although such heads have been known, described and patented from the 1940's onwards, their use is not widespread. They are mainly utilized for measuring and calibrating tapes and occasionally in digital recording. It appears that production of multi-track flux reading heads and electronics is more expensive than that of the conventional read heads.

Two types – the reluctance-variation and the Hall-effect types – are used to some extent [8, 9, 10, 34, 35, 36, 37, 38, 39]. The principle of the reluctance modulation head is explained in connection with Fig. 3.16(a). The flux lines

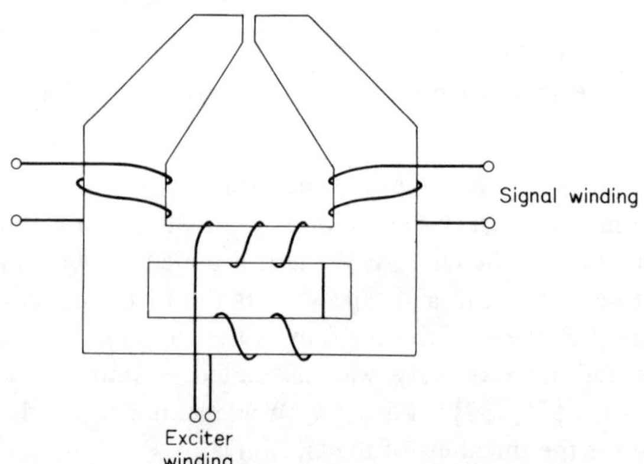

Fig. 3.16(a). Reluctance-modulation flux-sensitive head.

originating from the source to be measured, follow the high-permeability path of the magnetic material of the head. If a section of the head is periodically saturated by an external 'exciter flux' the reluctance for the original flux lines will be increased during the saturation period. Saturation of the thoroidal part of the head is equivalent to introducing a large gap in the original flux line path because in the saturated area $\mu = 1$. If a symmetrical exciter flux waveform is used and the core is saturated twice (once in positive, once in negative direction), the second harmonic of the exciter waveform, proportional to the flux to be measured, will appear at the signal coil terminals. The principle – known as magnetic modulator – has been extensively used in many areas of measuring magnetic fields, DC amplification, independently of magnetic recording applications.

One possible form of a transducer incorporating a Hall-effect device is illustrated in Fig. 3.16(b). The flux from the recorded tape is guided to the faces of the Hall-plate by the high-permeability pole pieces. This field deflects the

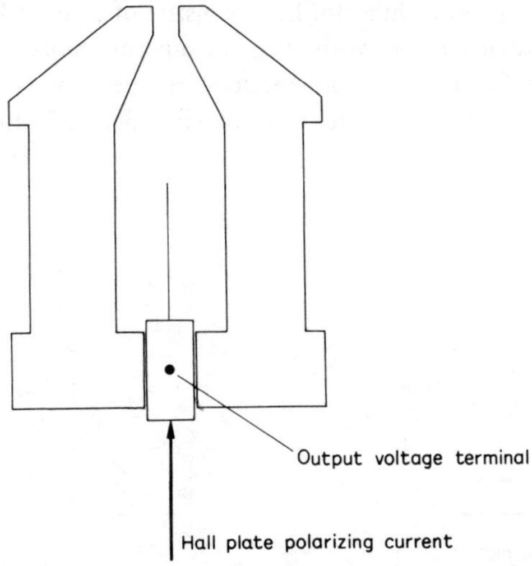

Fig. 3.16(b). Hall-effect flux-sensitive head.

polarizing current through the Hall-plate and generates the voltage across it. The output voltage for typical track widths and recording media is very low; further, many problems arise in construction of multi-track heads of this type. One advantage of Hall-effect heads over reluctance modulation type flux-sensitive heads is that they do not need a relatively high power high-frequency exciter source which may interfere with the field to be measured. However, because of their poor signal-to-noise ratio and low output, application of Hall-effect heads is very limited.

3.10. NOVEL HEAD CONSTRUCTIONS

The read/write heads which are in use even in the latest magnetic recording equipment are variants of the conventional ring head. Materials, manufacturing techniques and technology has greatly improved, tolerances were reduced and reliability increased but the construction is basically the same as in the 1930's: a high permeability core which has a narrow gap and carries the read/write coil. Multi-track heads are expensive pieces of precision engineering and many attempts are in progress to replace the ring head with alternative constructions which permit high recording density and lend themselves for inexpensive mass-production methods.

The use of thin Fe–Ni films which are either vacuum-deposited, plated or photo-etched seems one of the promising areas. Fabrication and analysis of such heads have been reported by L. T. Romankiv et al. [40], Y. Watanabe [41], D. Augier and J. P. Lazzari [42], J. P. Lazzari and I. Melnick [43], E. P. Valstyn and D. W. Kosy [44], A. D. Kaske et al. [45], Y. S. Lin and J. H. Tolaba [46].

In its simplest version a thin-film head consists of a single high-conductivity copper layer sandwiched between two high-permeability films. The high-permeability films form a horse-shoe-shaped magnetic yoke around the conductor and are perpendicular to the medium (Fig. 3.17). The equivalent of the

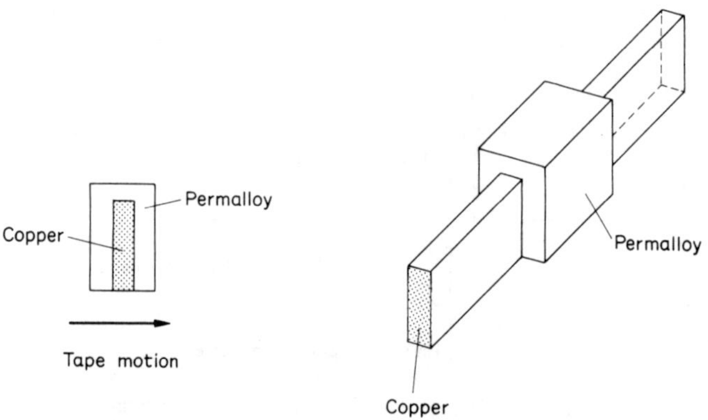

Fig. 3.17. Thin-film head schematic.

recording gap of the conventional head is now defined by the thickness of the copper conductor and the pole-tip length by the thickness of the high-permeability film. The ratio of pole-tip length to gap length is near unity whereas on conventional heads this ratio is large. [64]

An alternative thin-film head construction is illustrated in Fig. 3.18. The high-permeability thin-film layers are parallel with the recording medium. The fabrication of such heads consisting of a copper strip with 25 μm x 6 μm dimensions on which a 2.5 μm thick layer of Permalloy was electro-plated, was described by L. T. Romankiv [40]. The gap length was 2 μm. The analysis of the magnetic circuit indicates that when used as recording head, 200 mA current will generate 790 Oe field throughout a 1 μm thick recording medium [44, 46].

Flux-sensitive heads can be constructed with thin-film technology [45]. The head consists of a copper strip with a SiO insulating layer on top of which a thin (2000 Å) Fe–Ni film is deposited with its easy magnetic axis aligned with the copper strip. This copper strip will serve as the read line (Fig. 3.19). A further copper strip, orthogonal to the first one and called *sense line* is provided. The sense line carries a Fe–Ni layer in which a narrow gap, parallel with the sense line is formed. Conventional longitudinal writing is accomplished by energizing the sense line; the direction of motion of the recording medium is parallel with the read line. When used for reading the read line is energized by either a pulse or a

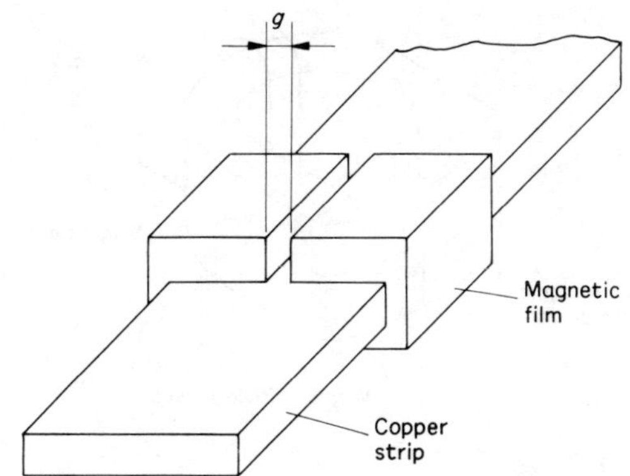

Fig. 3.18. Geometry of a thin-film head [39, 62].

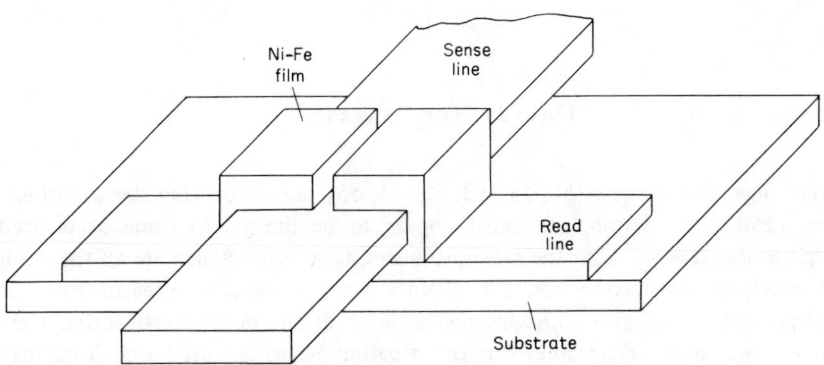

Fig. 3.19. Geometry of a flux-sensitive thin-film head [62].

sinusoidal voltage generating a hard-axis field in the head which the external field from the recorded medium steers from the hard-axis direction to a direction determined by the polarity of the field to be read. The external field which is required to steer the head magnetization is low, thus the structure is suitable for reading fields generated by very thin low-coercivity media.

A number of different head constructions which are closer to the conventional ring head but permit higher recording density were reported. W. T. Frost et al. [47] have achieved 20 000 flux reversals per inch on 2.9 mil wide tracks with a head where the recording gap is created by the junction point of two ferromagnetic pole-pieces (Fig. 3.20). Other constructions are based using the magnetic field of a single wire which carries heavy current and shaping the field by partially surrounding the wire with high permeability material. Instead of wire a thin film of good electrical conductor can be used [48, 49]. Many

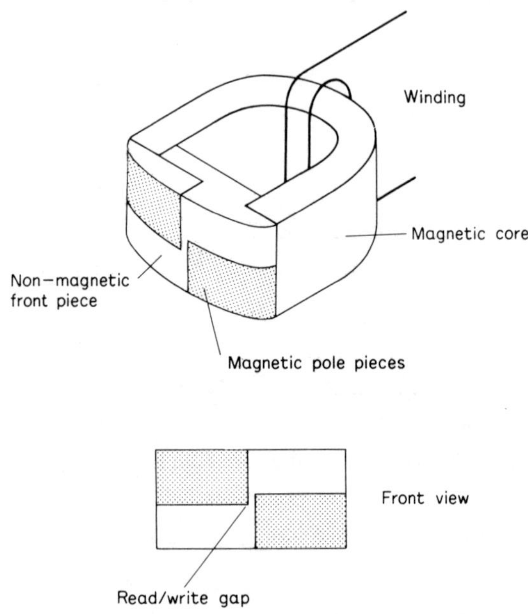

Fig. 3.20. Overlap head [47].

other ingenious designs [50, 51, 52, 53, 54, 55] are challenging the traditional ring head construction, but most appear to be limited to some specialized application. It seems that the most promising approaches which are on the point of reaching mass-production (as distinct from laboratory experiments) are hot-pressed ferrite and thin-film heads. Most of the novel construction techniques are more adaptable for the floating heads as used on discs than multi-track tape heads.

REFERENCES

1. Stewart, W. E. (1958), *Magnetic recording techniques*, McGraw-Hill.
2. Kornei, O. (1953), 'Structure and performance of magnetic transducer heads', *J. of Audio Eng. Soc.*, **1**, 3, 225–231.
3. Hoagland, A. S. (1958), 'High resolution magnetic recording structures', *IBM J. of R. & D.*, **2**, 2, 90–105.

4. Speedy, C. B. (1963), 'An analysis of the magnetic circuit of multi-track record and playback heads', *IEEE Trans. Audio,* **11**, 4, 127–133.

5. Baker, S. K. and Speedy, C. B. (1967), 'Crosstalk between in-line multi-track recording heads', *Proc. IEE,* **14**, 333.

6. Hoagland, A. S. (1956), 'Magnetic data recording theory: head design', *Communication and Electronics (AIEE),* **75**, 506–512.

7. Shew, L. F. (1962), 'High density magnetic head design for non-contact recording', *IRE Trans. Electronic Computers,* **EC-11**, 6, 764–773.

8. Kilburn, T., Hoffman, G. R. and Wolstenholme, P. (1956), 'Reading of magnetic records by reluctance modulation', *Proc. IEE (GB),* **103**, Part B, 333–336.

9. Edwards, D. G. B., Dunstan, E. M. and Tunis, C. J. (1964), 'The design of second harmonic detector heads', *The Radio and Electronic Engineer,* **27**, 99–116.

10. Sebestyen, L. G. (1964), 'The Honeywell flux sensitive head', *Honeywell Internal Report,* September.

11. Rettinger, M. (1955), 'Magnetic head wear investigations', *Journal of Soc. Motion Picture and Television Engineers,* **64**, 179–183.

12. Pear, C. B., Jun. (1970), 'A solution to tape recording head wear', *Computer Design,* **9**, 99–104.

13. Lufcy, C. W. and Heath, W. T. (1955), 'Alloy improves magnetic recording', *Electronics,* **28**, 137–139.

14. Nachman, F. J. and Buehler, W. J. (1954–1955), 'The fabrication and properties of Alfenol alloys Al–Fe containing 10–17 percent Al', *Navord Report 4130,* 1955. U.S. Naval Ordnance Lab.; also, *J. Appl. Phys.,* **25**, 3, 307–313.

15. Rose, K. H. and Glen, G. C. (1958), 'Recent developments in soft magnetic alloys', *Electrical Manufacturer,* **3**, 1.

16. Tebble, R. S. and Craik, D. G. (1969), *Soft Magnetic Metals and Alloys, Magnetic Materials,* Chapter 13, Wiley.

17. Chin, G. Y. (1971), 'Review of magnetic properties of Fe–Ni alloys', *IEEE Trans. Mag.,* **Mag-7**, 1, 102–113.

18. Westmijze, W. K. and Kleis, D. (1950), U.K. Patent No. 758 343.

19. Duinker, S. (1960), 'Durable high resolution ferrite transducer heads employing bonding glass spacers', *Philips Res. Reports,* **15**, 342–367.

20. Peloschek, H. P. and Vrolijks, M. H. M. (1964), 'Dense ferrites and the techniques of glass bonding for magnetic transducer heads', *Internat. Conference on Magnetic Recording, IEE, London,* 82–84.

21. Bakos, G. P. (1964), 'Design aspects of ferrite magnetic heads', *Internat. Conference on Magnetic Recording, IEE, London,* 85–87.

22. Hirota, E., Mihara, T., Ikeda, A. and Chiba, H. (1971), 'Hot pressed Mn–Zn ferrite for magnetic recording heads', *IEEE Trans. Mag.,* **Mag-7**, 3, 337–341.

23. Fisher, R. D. and Bladen, J. D. (1971), 'Single crystal manganese zinc ferrite recording heads', *IEEE Trans. Mag.,* **Mag-7**, 3, 350–351.

24. Sugaya, H. (1968), 'Newly developed hot pressed ferrite head', *IEEE Trans. Mag.*, **Mag-4**, 3, 295–301.

25. Finke, G. B. (1969), 'Efficiency of magnetic recorder heads', *IEEE Trans. Mag.*, **Mag-5**, 1, 47–50.

26. Hawkins, J. K. (1968), *Circuit design of digital computers*, J. Wiley & Sons, 447–448.

27. Begun, J. S. (1949), *Magnetic Recording*, Chapter 5, Murray Hill Books Inc.

28. Duinker, S. (1957), 'On the resolving power in the process of magnetic recording', *Tijdschr. Ned. Radiogenoot*, **22**, 29.

29. Frost, W. T. (1960), 'The sensitivity of reproducing heads', *IRE Wescon Convention Record*, **4**, 5, 46–49.

30. Norwood, R. E. (1969), 'Effects of bending stiffness in magnetic tape', *IBM J. of Res. and Dev.*, **13**, 2, 205–208.

31. Baumeister, H. K. and Nejezchleb, R. V., U.S. Patent No. 3 170 045.

32. Baumeister, H. K. (1963), 'Nominal clearance of foil bearing', *IBM J. of Res. and Dev.*, **7**, 153.

33. Weidenhammer, J. A. and Barbeau, R. A., U.S. Patent 3 327 916.

34. Kornei, O. (1965), 'Survey of flux-responsive magnetic reproducing heads', *J. Audio Eng. Soc.*, **2**, 3, 145.

35. Daniel, E. D. (1955), 'Flux sensitive reproducing head for magnetic recording systems', *J. Inst. E. E. Pt. B*, **102**, 4 (London), Paper 1856R.

36. Ferber, L. W. (1958), 'Flux responsive magnetic heads for low speed read out of data', *IRE Natl. Conv. Record Pt.*, **4**, 279.

37. Camras, M. (1962), 'Some experiments with magnetic playback using Hall effect sensitive elements', *IRE Natl. Conv. Record Pt.*, **7**, 80–83.

38. Uemura, S. (1958), 'Flux responsive reproducing head', *Electrotechnical Journal of Japan*, December, 142–135.

39. Kerr, D. and Quirk, J. M. (1960), 'A magnetic read head with output signal independent of tape speed', *J. Brit. IRE*, **20**, 743–748.

40. Romankiv, L. T., Croll, I. M. and Hatzakis, M. (1970), 'Batch-fabricated thin-film magnetic recording heads', *IEEE Trans. Mag.*, **Mag-6**, 3, 597–601.

41. Watanabe, Y. and Matsumoto, S. (1969). 'Fabrication of grouped magnetic heads', *IEEE Trans. Mag.*, **Mag-5**, 3, 451.

42. Augier, D. and Lazzari, J. (1971), 'Write process study of integrated magnetic heads', *IEEE Trans. Mag.*, **Mag-7**, 3, 679–683.

43. Lazzari, J. P. and Melnick, I. (1970), 'Recording integrated magnetic heads', *IEEE Trans. Mag.*, **Mag-3**, 3, 601–602.

44. Valstyn, E. P. and Kosy, D. W. (1969), 'The write field of a magnetic film recording head', *IEEE Trans. Mag.*, **Mag-5**, 3, 442–444.

45. Kaske, A. D., Oberg, P. E., Paul, M. C. and Sauter, G. F. (1971), 'Vapor deposited thin film recording heads', *IEEE Trans. Mag.*, **Mag-7**, 3, 675–679.

46. Lin, Y. S. and Toiaba, J. H. (1970), 'Analysis of write field distribution in permalloy recording head', *J. Appl. Phys.*, **41**, 1104.

47. Frost, W. T. and Neukermans, A. (1968), 'Narrow tracking magnetic recording head', *IEEE Trans. Mag.* **Mag-5**, 3, 445–448.

48. Barton, J. C. and Stockel, C. T. (1964), 'A novel type of magnetic recording head', *Radio and Electronic Eng.*, **27**, 11–18.

49. Truman, N. N. (1969), 'A low inductance strip-line recording head', *IEEE Trans. Mag.*, **Mag-5**, 1, 43–47.

50. Smaller, P. (1964), 'A flux-sensitive reproduce head of thin film construction', *Proc. of the Intermag. Conference, Washington, D.C., April.*

51. Kump, H. J. (1962), 'Single turn recording head', *Electronics*, **5**, 35.

52. Max, E. (1970), 'Thin film magnetic head', *IBM Tech. Disclosure Bull.*, **13**, 1, 248.

53. Haertlein, E. E. (1971), 'Multi element magnetic head', *IBM Tech. Disclosures*, **13**, 8, 2333.

54. Thornley, R. F. (1971), 'Laminated magnetic head', *IBM Tech. Disclosures*, **13**, 8, 2262.

55. Tchon, W. E. and Rodbell, D. S. (1970), 'A new magnetic recording head fabricated from coaxial wire', *IEEE Trans. Mag.*, **Mag-6**, 3, 593–597.

56. Smith, J. T. (1959), 'Design of a vertical magnetic recording head with large scale models', *Communication and Electronics*, **September**, 351–362.

57. Walther, G. L. (1964), 'Digital signal response of magnetic reproducing heads', *Philips Res. Rep.*, **19**, 281–295.

58. Fay, L. E. III (1965), 'Design criteria for magnetic recording heads', *IEEE Internat. Conv. Rec. USA*, **13**, 2, 189–194.

59. Peters, C. J. (1963), 'Fabrication of magnetic tape transducer by electroplating and engraving', *IRE Trans. Audio*, **Au-10**, 3, 79–83.

60. Walther, G. L. (1964), 'Reduction of crosstalk in multiple head structures for digital tape units', *Internat. Conference on Magnetic Recording, IEE, London,* 65–67.

61. Geurst, J. A. (1964), 'Suppression of crosstalk between magnetic heads in magnetic tape memory systems', *Internat. Conference on Magnetic Recording, IEE, London,* 62–64.

62. Kump, H. J. (1962), 'The magnetic configuration of stylus recording', *IRE Trans. Electronic Computers*, **EC11**, 2, 263–273.

63. Hughes, G. F. (1971), 'On digital head design', *IEEE Trans. Mag.*, **Mag-7**, 3, 695–699.

64. Potter, R. I., Schmulian, R. J. and Hartman, K. (1971). 'Fringe field and readback voltage computation for finite pole tip length recording heads', *IEEE Trans Mag.*, **Mag-7**, 3, 689–695.

65. Hodder, W. K. and Monson, J. E. (1971), 'Field analysis for magnetic heads with eddy currents or complex permeability', *IEEE Trans. Mag.*, **Mag-7**, 3, 686–689.

66. Lederie, G. M. (1971), 'Heat transfer calculations at the tape-head interface of a computer tape drive', *IBM J. Res. and Dev.*, **15**, 236–241.

67. Fay, L. E. (1964), 'Two dimensional theory of the recording head', *Internat. Conference on Magnetic Recording, IEE, London,* 71–73.

68. Clark, D. L. and Merril, L. L. (1947), 'Field measurements on magnetic recording heads', *Proc. IRE.*, **35**, 1575–1579.

69. Fischer, R. B., Cox, C. M. and Holdinsky, C. (1970), 'Computer aided testing and fabrication of magnetic tape heads', *IBM J. Res. and Dev.*, **14**, 633–640.
70. Davies, A. V. (1966), 'Causes of head-to-medium separation', *IEEE Trans. Mag.*, **Mag-2**, 782–783.

4 Recording theory

4.1. INTRODUCTION

Fundamental contributions to the modern theory of recording and reproduction of sinusoidal waveforms were made independently by R. L. Wallace [1] and W. K. Westmijze [2] whose work has formed the foundation of many subsequent treatments of the subject. Wallace computed the field corresponding to sinusoidal magnetization distribution in the recording surface, and with the method of images, calculated the change introduced by the presence of the high permeability reproduce head. Westmijze, in an extensive study of magnetic recording and reproduction, calculated the magnetic field in the vicinity of the record head gap and then by applying the reciprocity theorem, determined the readback flux through the read head coils. Both studies were primarily concerned with analogue recording and sinusoidal voltage. Important contributions to the theory of digital recording were made subsequently by D. F. Eldridge [3], A. S. Hoagland [10], K. Teer [4], J. J. Miyata and R. R. Hartel [7], S. Duinker [26] and I. Stein [6].

The traditional analysis of a saturation recording process starts with a simplified model of the system (infinite permeability for the head material, unity permeability for the medium, uniform saturation throughout the medium, step-function reversal of magnetization, longitudinal record head field only, negligible self-demagnetization) and then refines the first-order approximations by correction factors for finite head permeability, gap length, demagnetization and non-ideal magnetization transition. A novel approach by S. Iwasaki and T. Suzuki [43] took into account the interaction of the self-demagnetizing field of the elementary permanent magnet generated by the recording process with the head field. This interactive model was extended and refined by N. Curland and D. E. Speliotis [17, 56]. The study by R. I. Potter and R. J. Schmulian [54] included the effect of the head motion, record current rise time and the reduction of demagnetizing field caused by the presence of the high-permeability head structure. A different method of analysis was used in a series of studies by B. Kostyshyn [42, 44, 55] based on resolving the stored magnetization pattern to its Fourier components.

The development of metal particle and metal film surfaces which, contrary to the traditional gamma ferric oxide, permit a wide variation of the magnetic parameters necessitated to review the theory in respect of dependence of recording parameters (pulse amplitude and width) on magnetic and geometric parameters of the medium (coercivity, remanence and thickness). Theories of recording on such media were presented by P. I. Bonyhard et al. [24], J. R. Morrison [21], A. V. Davies et al. [19], M. Nishikawa [20], B. K. Middleton [45]. An extensive review of the present state of recording theory, accomplishments and unresolved problems was given by D. E. Speliotis [28, 29] and R. O. McCary [12].

The present state of understanding of the processes which take place in saturation digital recording and reproduction is fairly good in respect of the effect of geometric parameters of the system such as the effect of gap length, gap depth, head permeability, medium thickness, head-to-medium separation, but less comprehensive in respect of the influence of magnetic parameters – coercivity, remanent intensity of magnetization – on the maximum recording density. It seems to be necessary to distinguish between 'thin' and 'thick' media, particulate and thin film surfaces which require different optimization of the magnetic parameters for optimum recording performance. D. S. Speliotis and J. R. Morrison [46] analysed the effect of remanence, coercivity and medium thickness on the width and amplitude of the reproduce pulse, and derived formulas for the relation of magnetic and recording parameters in thin films.

In this chapter we will review the write process, the self-demagnetization, the read process, the phenomena of pulse asymmetry, pulse crowding and peak shift, and derive expressions for the amplitude and width of the reproduced pulse as a function of the parameters of the medium and the magnetic head.

4.2. THE FIELD OF THE RECORDING HEAD

In our preliminary description of magnetic-surface recording we referred to the recording process as the magnetization of a minute area of the recording medium. The air gap in the recording head is brought into close proximity with the medium and when the head is driven by an nI magnetomotive force, the leakage flux in the vicinity of the gap saturates the thin layer of magnetizable material. Hence when the medium moves away from the head field it retains a remanent magnetization determined by head field, the magnetic parameters of the medium and its self-demagnetization.

The model which we used for computing the field in the vicinity of the recording head gap in Fig. 3.9 is a valid representation of many recording situations hence the results obtained by using it are of general interest. The starting point was the observation that the field in the vicinity of the gap can be expressed as

$$H = H_g \frac{g}{l_a},$$

(4.1)

in which H_g is the homogeneous field in the gap, g the gap length, l_a the length of the flux path in the air. As l_a is a semi-circle we can substitute $l_a = r\pi$, or

$$H = \frac{H_g}{\pi\sqrt{x_n^2 + y_n^2}} \tag{4.2}$$

Equation (4.2) can be considered as a good approximation for the area remote from the gap, but breaks down in the immediate vicinity of the gap because, when $x_n = y_n \to 0$, $H \to \infty$. Although the geometry of the idealized head is quite simple, a precise analytic solution is not obtainable in closed form. The approximation which has been found to give satisfactory accuracy and has been used in many studies of magnetic recording is the Karlquist equation [36] :

$$H_x(x, y) = \frac{H_g}{\pi}\left(\tan^{-1}\frac{g/2 + x}{y} + \tan^{-1}\frac{g/2 - x}{y}\right) \tag{4.3}$$

$$H_y(x, y) = -\frac{H_g}{2\pi}\ln\frac{(x + g/2)^2 + y^2}{(x - g/2)^2 + y^2} \tag{4.4}$$

where H_g is the homogeneous field inside the gap.* The maximum longitudinal field will occur in the centre line of the gap ($x = 0$) and is:

$$H_x(0, y) = \frac{2}{\pi} H_g \tan^{-1}\frac{g}{2y}$$

and for the remote field ($2y \gg g$, $H_g \sim 1/g$), Equation (4.5) reduces to

$$H_x \cong \frac{1}{\pi y}. \tag{4.6}$$

The more accurate expression for the head field derived by C. J. Fan [11] differs from the Karlquist equation at points close to the edge of the gap only; at other points the two are in good agreement [48].

Equations (4.3) and (4.4) represent circles with their centres located on the x and y axis, respectively, as illustrated in Fig. 4.1.† The H_x circles connect those points on the plane along which the x component of the field is constant. It is noticeable that the peak of the y field at any distance is less than the peak x field and that the y field is less concentrated than the x field.

4.3. MEDIUM MAGNETIZATION

Having established the spatial distribution of the head field, a far more complex question has to be answered: What is the resultant magnetization in a medium upon which this field is impressed? We will start the investigation with a simplified model, assuming that the field has been sufficiently strong to saturate

---

* Depending whether the field deep in the gap or the field at the edge of the gap is used as reference, the equations differ by a factor of 2.

† Alternative representation of the field was shown in Figs. 3.11 and 3.12.

the full cross-section of the thin medium in one direction, and the write current reversal changes the medium saturation in opposite direction. This saturation-to-saturation* flux reversal is considered 'nearly instantaneous', i.e. we neglect the effect of the head inductance and capacitance on the current rise time and

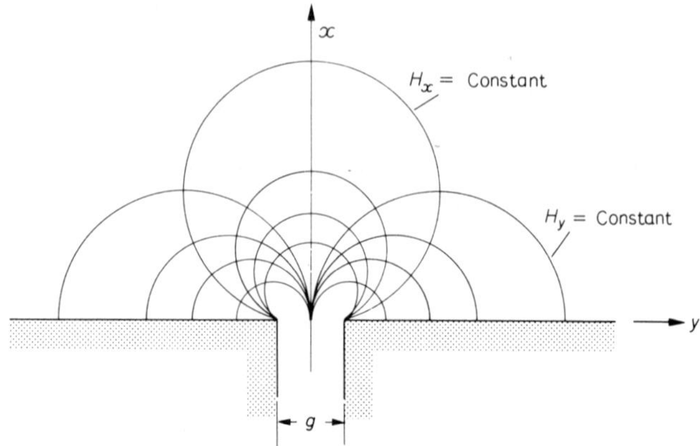

Fig. 4.1. Constant field contours in the vicinity of the recording gap.

similarly ignore the time-delay in the build-up of the magnetic field caused by eddy-current losses. Only the x component of the field will be considered. The model is fairly representative of the actual situation and as such is of general interest. The neglected effects will appear as correction factors of the simple model.

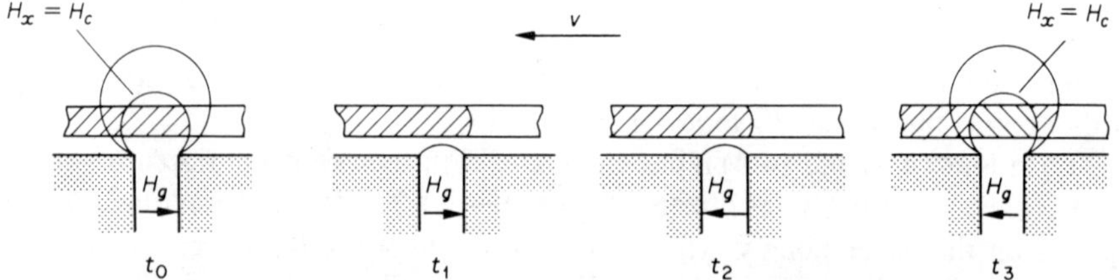

Fig. 4.2. Schematic representation of saturation recording.

The field representation in Fig. 4.2 is a useful aid for visualizing the sequence of events. Let us start the observation at the time t_0 when the head field is at its peak level. The area into which the head field extends can be divided into distinct zones: the 'high-field' zone where the field strength exceeds the coercivity of the medium, and 'low-field' zone where the field is below the coercive force, and between the two an area where the head field is sufficiently strong to disturb the magnetization of the medium but does not saturate it in

* The term 'saturation' is used here to describe the case when the peak magnetization of the medium is approximately equal to the remanence of its major hysteresis loop.

the new direction. The tape moves with v velocity in the direction indicated in Fig. 4.2. The medium to the left of the gap and inside the H_c circle is shown as saturated in the direction of the H_g gap field. At some point in time the head current is turned off; at time t_1 it is nearly zero. The medium is left with a magnetization corresponding to B_r (self-demagnetization is ignored). At time t_2 and head current and field starts to increase in the opposite direction and at t_3 it reaches its full final value. During t_0 to t_3 time, the medium movement was very small At time t_3 the material inside the $H_x = H_c$ circle is saturated in the opposite direction as it was at t_0. After the head field changed polarity the medium moving out of the head field will retain a magnetization experienced at the trailing edge of the gap near the $H_x = H_c$ circle.

Admittedly this picture of the recording process is in many respects inaccurate but it permits the establishment of a number of qualitative conclusions, well proven by experiments. It indicates that the transition from one saturated state to the opposite and the corresponding peak of the reproduced voltage, will not be in the centre of the gap but in the direction of the trailing edge. Increasing the recording current will move the transition further away from the centre, past the trailing edge. The transition, which is the boundary between two permanent magnets in the schematic representation of recorded tape, must have a finite width. This transition area lies somewhere outside the $H_x = H_c$ circle. The width of this transition will be primarily determined by the gradient of the head field and the slope of the sides of the hysteresis loop. The analytical approach to the problem of establishing the magnetization in the medium can be based on approximating the hysteresis curve of the material with a conveniently chosen function. Many such functions of various degree of complexity can be found.* Now using the Karlquist equation, or some other approximation of the head field, the magnetization in the medium for a single transition can be computed as a function of the medium's saturation, remanence, coercivity and the ratio of the applied head field to the saturation field [3, 4, 5, 6, 7, 9, 54, 75 (pp. 460–474)].

The model of 4.2 has not taken into consideration the self-demagnetization of the recorded transition. Self-demagnetization is thought to be the factor which ultimately limits the recording density hence it is of principal significance. We make a short digression here to recapitulate the essential facts.

4.4. SELF-DEMAGNETIZATION

When a ferromagnetic material is placed in an H_a magnetic field the magnetized material generates an H_d field which is in opposition to the applied field. Consider the ring magnet of Fig. 4.3 which has two segments with different

* See references [5, 44, 54, 75] for examples.

permeabilities and cross-sections. In each segment the permeability μ and cross-section A are constant; the flux remains confined to the material, i.e. there is no leakage. For this case

$$\oint H\,\mathrm{d}l = H_1 l_1 + H_2 l_2 = \frac{4\pi}{10}nI,\qquad(4.7)$$

and

$$B_1 = \mu_1 H_1 = \frac{\phi}{A_1}.\qquad(4.8)$$

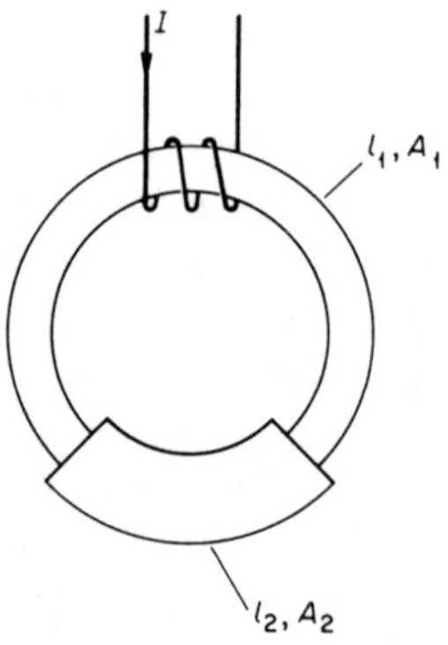

Fig. 4.3. Ring magnet.

Similarly

$$B_2 = \mu_2 H_2 = \frac{\phi}{A_2},\qquad(4.9)$$

$$H_2 = H_1 \frac{A_1 \mu_1}{A_2 \mu_2}.\qquad(4.10)$$

From Equations (4.7) to (4.10) the solution for B_1 is obtained in the form:

$$B_1 = \frac{4\pi}{10}\frac{nI}{l_1}\mu_1 - \frac{A_1 \mu_1^2 l_2}{A_2 \mu_2 l_1}H_1$$

$$= \alpha I - \mu_d H_1,\qquad(4.11)$$

where μ_d is the demagnetization factor.

In general B_1 is a nonlinear function of H_1 and it may be represented as $B_1 = f(H_1)$. The solution for B_1 must satisfy simultaneously $B_1 = f(H_1)$ and $B_1 = \alpha I - \mu_d H_1$ equations. Analytical solutions necessitate to find an approximation for $B_1 = f(H_1)$ by a function which fits the actual hysteresis curve for the material for at least a limited region; such functions can be generated; however the B–H relation of the material is nearly always given in graphical form and it is more convenient to use graphical methods. The graphical solution is illustrated in Fig. 4.4: the intersection of the $B_1 = f(H_1)$ hysteresis curve with $-\mu_d H_1$ gives the point to which the magnetization will relax after the magnetomotive force was removed. The second quadrant of the B–H or M–H loop which shows the magnetization as the magnetizing field is reversed is often termed as the 'demagnetization curve' of the material.

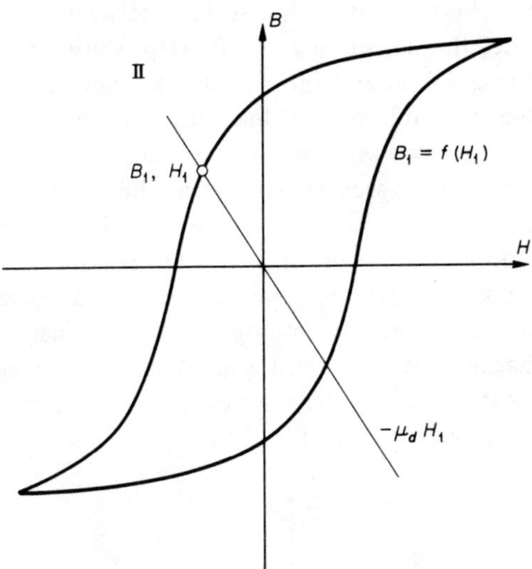

Fig. 4.4. Graphical determination of demagnetization.

Returning now to magnetic recording we want to consider the effect of self-demagnetization on the isolated saturation-to-saturation flux reversal and on the elementary permanent magnet generated by two closely spaced flux reversals. The slope of the written transition before demagnetization is equal to the product of the head field gradient $\partial H/\partial x$ and the slope of the demagnetization curve:

$$\frac{\partial M}{\partial x} = \frac{\partial H}{\partial x} \times \frac{\partial M}{\partial H}.$$

Clearly a narrow transition requires a material with a 'square' hysteresis loop and a sharp gradient of the trailing leg of the head field. The widening of the written transition can be qualitatively explained with the aid of our Fig. 4.2. Particles of the medium which are just outside the saturation zone of the head field and are moving away from the gap centre line when the field is reversed do not experience the saturation field, thus they end up unsaturated causing the widening of the transition zone. This effect is often referred to in the literature of magnetic recording as 'recording demagnetization'.

A length of saturated surface between two transition can be considered as a bar magnet. Computation of the residual magnetization for the general case when the external flux path is not confined to a narrow air gap as it was in our example of Fig. 4.3, is rather involved [2, 33, 39, 75]. The well-known general rule is that the demagnetization factor is primarily a function of the length-to-diameter ratio of the magnet; it is negligible for long bar magnets but significant in high-density saturation recording where the length of the elementary magnet generated by the recording process is comparable with the thickness of the medium. At the highest densities, the surface between two transitions is not saturated any more, the transition zones overlap. During the recording process when the tape is magnetized and the recorded magnetization short-circuited by the high-permeability recording head, the situation can be approximated by our example of the ring of two segments in Fig. 4.3. One segment is the recorded permanent magnet, a short section with permeability of near-unity because of saturation, the other sector is the high-permeability head. There will be a partial demagnetization during and immediately after recording; however, the full demagnetization takes place after the recorded area moves away from the high-permeability head. During recording, the tape is magnetized to B_s. After removal of the magnetizing field and the influence of the magnetic short-circuit of the high permeability head, the magnetization will relax along the B_r–B_m path to the point indicated by B_m (Fig. 4.5).

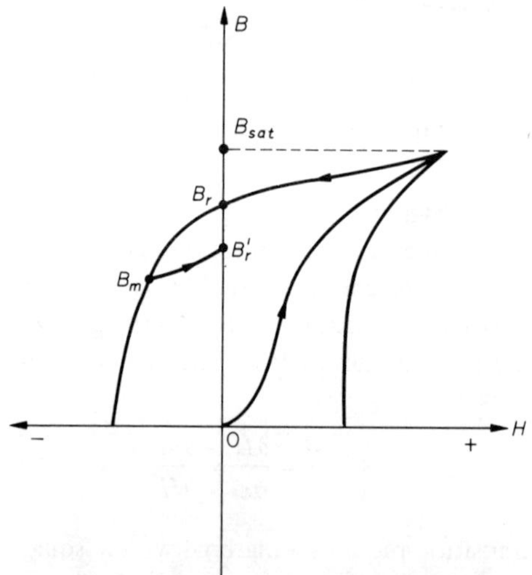

Fig. 4.5. Illustration of demagnetization and partial recovery on the hysteresis loop.

During the reproduction process, the demagnetizing field is again shunted by the high permeability read head; the induction increases along $B_m - B_r'$ path (Fig. 4.5). There is an irreversible loss in magnetization represented by $B_r - B_r'$.

Figure 4.6 illustrates the effect of increase in flux density and coercivity on the remanent magnetization on an idealized hysteresis loop. In presence of strong demagnetizing field it is the increase in coercivity which has a greater effect on remanent magnetization.

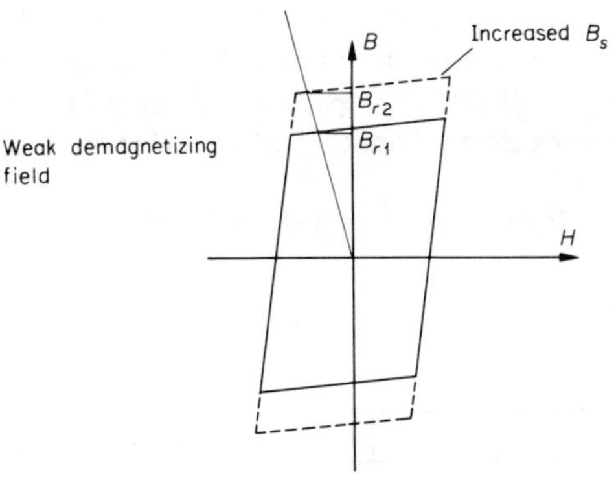

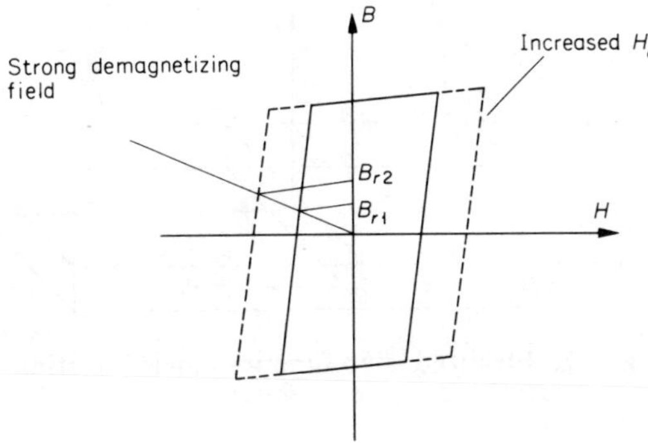

Fig. 4.6. Effect of increased B_r and H_c on remanent magnetization in case of weak and strong demagnetizing fields.

4.5. MAGNETIZATION TRANSITION

Our principal concern with respect to self-demagnetization, is not so much the loss of amplitude, but the effect it has on maximum recording density. Let us consider an isolated saturation-to-saturation transition as in Fig. 4.7. This ideal transition or 'step function' transition is of fundamental interest as any pattern of magnetization changes made up of alternations between two discrete states can be resolved into a series of step functions. Clearly, the magnetic surface cannot support such an abrupt transition and there will be a finite transition region. Our interest is the width of the transition length and the effect of demagnetization on this transition length. The idealized step-function transition may be approached by an extremely thin recording medium with perfectly square hysteresis loop. For the idealized conditions D. E. Speliotis and J. R. Morrison [9] following the approach of J. J. Miyata and R. R. Hartel [7] concluded that for a medium which possesses M_r remanent magnetic moment per unit volume (remanent magnetization) directed along the x axis so that it points in the $+x$ and $-x$ directions respectively at the pole sheet, the maximum demagnetizing field along the medium centre line is (Fig. 4.7)

$$H_x(x_0, 0) = H_d = 8M_r \tan^{-1} \frac{t_m}{2x_0} \qquad (4.12)$$

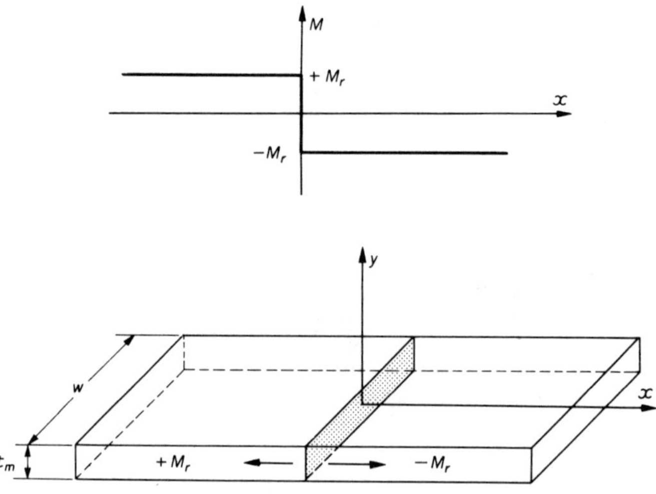

Fig. 4.7. Idealized 'step function' magnetization.

The problem was treated as two-dimensional with t_m medium thickness. Making the assumption, following D. W. Chapman [66] that the maximum value of the demagnetizing field is equal to the coercivity of the recording medium (our Fig. 4.3 which we used to introduce the concept of self-demagnetization clearly indicates that the demagnetizing field cannot exceed the coercivity of the material) the minimum width of the transition is calculated as

$$2x_0 = t_m \frac{1}{\tan \dfrac{H_c}{8M_r}}. \qquad (4.13)$$

Equation (4.13) is an expression for the transition width as function of thickness, coercivity and remanent magnetization of the medium caused by self-demagnetization.

A more realistic model of the actual situation assumes a transition region in which the magnetization varies according to some well-defined law. A uniform distribution over a length l_0 is an obvious assumption (Fig. 4.8). The demagne-

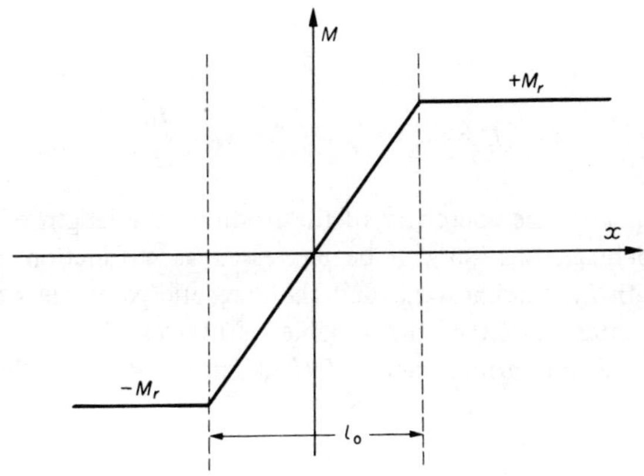

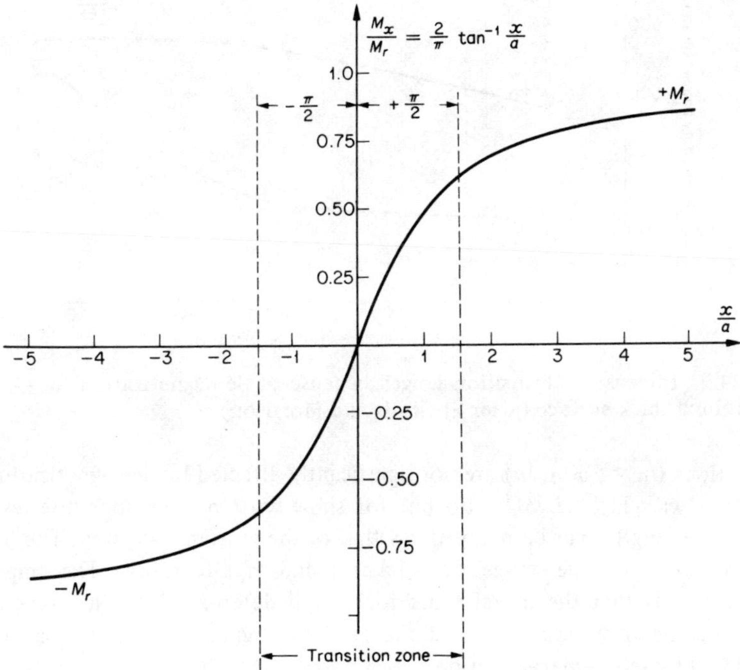

Fig. 4.8. Ramp and arctangent magnetization.

tizing field for this case in the centre plane of the medium is after D. E. Speliotis and J. R. Morrison [9].

$$H_x(x_0, 0) = H_d = \frac{2M_r}{l_0}\left[t_m \log \frac{D^2 + t_m^2/4}{C^2 + t_m^2/4} + 4 \ \ D \tan^{-1}\frac{t_m}{2D} - C \tan\frac{t_m}{2C}\right].$$

$$(4.14)$$

Here

$$D = x_0 + \frac{l_0}{2}, \qquad C = x_0 - \frac{l_0}{2}.$$

If H_x is set equal to the coercivity of the medium, the length of the transition region after demagnetization can be expressed as a function of the written transition length l_0, thickness t_m and the magnetic parameters of the surface. The complex interrelation of the variables is illustrated for a $t_m = 500 \ \mu$ in recording material for various values of H_c/M_r in Fig. 4.9. Relatively long initial

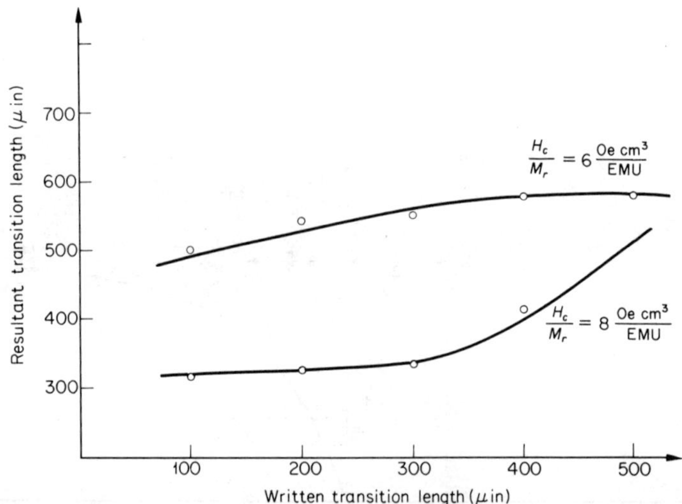

Fig. 4.9. Increase of transition length because of demagnetization for $t_m = 500$ microinch thick surface (after Speliotis and Morrison).

transitions ($l_0 = 500 \ \mu$ in) are not significantly effected by demagnetization in a material with high H_c/M_r ratio but for short written transitions, the resultant transition length may be many times that of the written transition. The trend is similar over a wide range of thickness and H_c/M_r ratios. The important conclusion is that the initial transition length determined by the slope of the sides of the hysteresis loop and the head field gradient may be significantly modified by self-demagnetization.

The linear distribution of magnetization in the transition region is only one of the many possible assumptions. J. J. Miyata and R. R. Hartel [7] have postulated an

$$M_x = -\frac{2}{\pi} M_r \tan^{-1} \frac{x}{a}, \tag{4.15}$$

distribution in which M_x is the longitudinal component of the magnetization, M_r the remanent magnetization and a, a parameter which controls the sharpness of the transition. The step function transition appears as a special case of the arctangent distribution when a approaches zero.

The length of the transition zone for an arctangent magnetization distribution is not sharply delimited and needs definition. One possible approach is illustrated in Fig. 4.8. The arctangent function is approximated by its tangent at the origin which has a slope $2/\pi$ and this line is extended to saturation ($M_x/M_r = 1$). The length of the transition zone is determined as $l_0 = 2 \times \pi/2$. The arctangent distribution has been chosen in many studies of the demagnetization because it asymtotically approaches a finite non-zero value, is an even function, and has a continuous derivative. [9, 15, 22, 24, 33, 45, 61, 62, 66]. The maximum value of self-demagnetization will now appear as the function of the medium thickness t_m, H_c/M_r, and the a parameter. Using various computation techniques and approximations, D. W. Chapman [66], J. J. Miyata and R. R. Hartel [7], D. E. Speliotis and J. R. Morrison [9], P. I. Bonyhard et al. [24], and R. I. Potter [15] calculated the demagnetizing field and minimum transition length considering demagnetization only. The parameter a can be approximated by

$$a = \frac{t_m}{2}\left(1 + \frac{1}{\left(\tan\dfrac{H_c}{4M_r}\right)^2}\right)^{1/2}. \tag{4.16}$$

For thin metal films where $H_c < 4M_r$, Equation (4.16) reduces to:

$$a \simeq 2t_m \frac{M_r}{H_c}, \tag{4.17}$$

and the minimum transition length will be

$$l_{\min} \cong 2\pi t_m \frac{M_r}{H_c} \tag{4.18}$$

The value $1/l_{\min}$ is often quoted in the literature of magnetic recording as the 'theoretical maximum recording density' which a recording medium, specified by its M_r/H_c ratio and thickness can support. It was derived for thin media considering the self-demagnetization as the ultimate density limiting factor. Clearly, the maximum density is, to some extent, a function of such parameters as recording head gap length, write current magnitude, and waveform. A detailed theoretical and experimental analysis of the recording demagnetization by J. R.

Morrison [33] takes into consideration the effect of the recording gap length, write current, and corresponding head field contours. Two significant factors emerge: (a) the recording demagnetization is slightly dependent on the recording gap length, and (b) there is a nearly linear relation between write current and demagnetization. Thin-film and relatively thick oxide media exhibit different demagnetization characteristics and in particulate media demagnetization is a function of particle orientation which takes place during the manufacturing process. However, Equation (4.18) is a useful approximation of the minimum transition length for thin-film media.

4.6. THE ITERATIVE HYSTERETIC MODEL

The studies on magnetization transition which we referred to in Section 4.5 have followed the pattern of establishing the initial magnetization distribution in the medium from the head field and medium hysteresis curve. This distribution was then modified by the self-demagnetization following the removal of the medium from the recording field and assuming that the maximum value of the demagnetizing is equal to the coercivity. The writing process and demagnetization appeared to be divorced from each other. The model of S. Iwasaki and T. Suzuki [43] postulates an initial magnetization distribution which is the resultant of the head field and demagnetizing field and not the head field alone. The effective total field which acts on the medium when the head field H_h is applied is the sum of the head field and the demagnetizing field H_d:

$$H = H_h + H_d. \tag{4.19}$$

The magnetization distribution which produces the H_d demagnetizing field is determined by the sum of H_d and H_h. Obviously an iterative process is required, starting with a magnetization distribution which is determined by the head field alone, then summing the head field and demagnetizing field to arrive at the corrected head field and repeating the process through several steps. The Karlquist approximation is used for the head field H_h and the demagnetizing field is computed (after W. F. Brown: [82], see also [17]) from

$$H(xyz) = \int_V \frac{\rho^*}{r^2} r^0 \, \mathrm{d}V, \tag{4.20}$$

which gives the magnetic field intensity at some point xyz due to volume charge distribution ρ^*. Here r^0 is the unit vector pointing from the volume element $\mathrm{d}V$ to the point xyz and r is the distance from $\mathrm{d}V$ to xyz. The simplification is made to consider the problem one-dimensional only, i.e. independent of z and y. The integration is now over the x variable only and this is approximated by sum of integrals over small regions between x_i and x_{i+1} so that the ρ^* is constant over each region (this is equivalent to approximate the magnetization distribution by a piece-wise linear function). To obtain convergence of the iterative process, S. Iwasaki and T. Suzuki have chosen the recursive formula:

$$M_x{}^n = M_x + \beta(M_x - M_x{}^{n-1}), \qquad (4.21)$$

according to which the $(n-1)$th magnetization $M_x{}^{n-1}$ is modified to the (n)th magnetization by M_x which was obtained by summing the H_h head field and the $H_d{}^{n-1}$ demagnetizing field. The latter was produced by $M_x{}^{n-1}$. A convenient value was chosen for the β convergence factor.

The computed magnetization patterns for the iterative model shows significant deviation from the magnetization computed from the head field alone, particularly in the close vicinity of the gap; further, they indicate that higher orientation in a particulate medium (which is equivalent of increasing the squareness ratio) reduces the maximum bit density. Experiments with gamma Fe_2O_3 tapes, keeping H_c almost constant and varying the squareness by changing the particle orientation seem to support the calculations.*

The iterative model was used in two studies of the magnetization distribution in the transition region between two oppositely magnetized regions by N. Curland and D. E. Speliotis [17, 56]. The first of the two follows closely the Iwasaki–Suzuki method and extends the same by a second set of computation for the condition which applies when the head field has been removed. Comparing the calculated transition lengths from various theories it appears that the iterative model yields always (i.e. for all values of H_c, B_r and t_m) a higher transition length than other theories. However, the second of the two studies which takes into consideration the effect of the high-permeability write core material on the recorded transition region corrects the first set of results to a lower value. The results indicate, that, at least for the case of a 0.25 μm thick, H_c = 400 Oe and squareness ratio S = 0.8 metal film for which the remanence was varied at 7 : 1 ratio the length of the transition region is shortest at low remanence, and the amplitude of the reproduce pulse increases nearly linearly with increase in remanence.

The iterative model was further extended by R. I. Potter and R. J. Schmulian [54] in a comprehensive study of the recorded magnetization. The model takes into consideration the reduction of the demagnetizing field by the presence of the high-permeability write head, the removal of the write head and subsequent re-magnetization by the read head. The readback voltage was calculated for the isolated transition with different recording current waveforms and amplitudes. Numerical computations were carried out for 1 μm read/write head gap length and t_m = 0.1 μm thick medium which had H_c = 300 Oe coercivity, M_r = 800 EMU/cm^3 remanence and, S = 0.9 and 0.8 squareness ratios. The current reversal took place while the head moved 0.25 μm distance and current reversal waveform was either ramp or exponential. The primary effect of lengthening the distance over which current reverses is to displace the transition, although some

---

* This is in conflict with other computations on the effect of squareness on maximum bit density which indicate the desirability of a 'square' hysteresis loop.

broadening of the readback pulse occurs. Decreasing squareness ratio increased the pulse width. The computations indicate that although there is a high degree of asymmetry in sequentially written pulses with respect to both amplitude and peak shift, a usable output voltage and resolution may be obtainable even up to 15 000 bits/inch density.

The complexity of the recording and self-demagnetizing process does not permit a description in a closed analytical form by a set of equations. Care should be exercised in extending the results of numerical computations which we quoted and which apply primarily to thin films to the more general case of relatively thick particulate media. The validity of the models on which the calculations were based are limited to a narrow range of the parameters involved.

4.7. LOCATION OF THE RECORDED TRANSITION

The model of the recording process in Fig. 4.2 indicates that the location of the recorded transition is beyond the trailing edge of the recorded gap. Increasing the recording current will move the H_c circle further away from the gap hence it is expected that the position of the recorded transition will move away from the gap centre line with increasing current. The problem of analytically determining the location of the recorded transition as a function of the medium parameters and system geometry is straightforward but rather cumbersome. Such computations were carried out by M. F. Barkouki and I. Stein [5] who calculated and measured the location of the reproduce pulse as function of the applied field and head-to-medium spacing for a specified tape. The effect of recording current on pulse location is very clearly illustrated by the ingenious magnified model of the head and medium by D. L. A. Tjaden [14, 16]. Increasing the recording current by a factor of four moved the location of the transition by about 1.5 gap length distance. The measurements of D. F. Eldridge [3] on a practical system with a 0.5 mil write and 0.25 mil read head gap have shown a total change in reproduce pulse location slightly in excess of 0.55 mil as the recording current was varied from 7 to 50 mA. The calculations and experiments clearly indicate that in high-density multi-track recording systems the scatter in recording head current and head sensitivity has to be kept within tight tolerances. Head current variation in a multi-track can cause an apparent skew in the same manner as mechanical gap scatter.

4.8. MEASUREMENT OF THE TRANSITION LENGTH

Direct measurement of the length and shape of the transition region between two oppositely saturated areas represents a difficult problem. The traditional technique of making the magnetized areas visible by coating or spraying with finely dispersed iron powder (Bitter technique) cannot be used because of the dimensions involved; the same applies to all other standard magnetic field measuring methods. Hence nearly all studies on the transition region rely on measurement of the reproduce pulse shape and width for verification of the theories and computation on the transition length. The interpretation of the results and discounting the effects of the reproduce head material, gap length, head-to-tape separation is by no means straightforward, thus some doubt remains when computed transition length values are compared through the width of the reproduced pulse. One method which has been successfully applied to direct observation of the recorded magnetization pattern is Lorentz transmission microscopy [22]. This method relies on deflection of an electron beam by the magnetization in the medium.

Iron films evaporated on glass substrates, pre-coated with a thin carbon film were prepared for electron microscopy observation. The film thickness was in the range of 200–600 Å and their coercivity varied from 50 to 400 Oe with a squareness ratio of 0.9 to 0.95. The films were recorded at 200 flux reversals per centimeter, floated off the substrate and mounted on 200-mesh grids.

The agreement between transition length computed with the iterative model and the measured length show a very good agreement over a narrow range (100 to 200 Oe) of coercivities; the difference increases at higher values.

4.9. THE REPRODUCTION PROCESS

The magnetization impressed on the tape during recording is reproduced by passing the tape over a read head which is generally similar in construction to the recording head. The readback voltage at the terminals of the read head will be proportional to the flux entering the head:

$$e = -n \frac{\mathrm{d}\phi}{\mathrm{d}t} = -n \frac{\mathrm{d}\phi}{\mathrm{d}x} \cdot \frac{\mathrm{d}x}{\mathrm{d}t} = -nv \frac{\mathrm{d}\phi}{\mathrm{d}x}. \tag{4.22}$$

Here v is the velocity of the recorded surface relative to the reproduce head, n the number of turns, x distance along the tape.

Our prime objective is to establish the width and magnitude of the pulse which is generated by reading an isolated saturation-to-saturation transition. Although pulse width and amplitude are not the only factors which determine the system performance – other factors as peak shift, noise, etc. have to be considered – but they are certainly the most important ones.

Two approaches are widely used in the analysis of the reproduction process: the method of images, following the calculations of R. L. Wallace [1] and the application of the reciprocity theorem [2, 3, 10, 59]. We will use here the reciprocity principle because of the insight it gives into the reproduction mechanism. The flux which emanates from the recorded magnetization and couples into the read head is very low; it is justified to proceed on the basis that the read head will act as a linear element. Also, the permeability of the saturated tape approaches unity and the head permeability can be considered infinite. Under those conditions the principle of reciprocity can be utilized to determine the flux which couples into the read head from the recorded magnetization.

During the analysis of the recording process various approximations were derived for the field distribution in the vicinity of the recording gap. Once this recording field distribution has been established, the flux ϕ through an arbitrary cross-section of the recording medium caused by a current i through the write head coil can be calculated. The reciprocity theorem states that this same i current encircling a cross-section of the medium excites the same flux in the read head coil. If the recording surface is replaced by a set of current sources of appropriate strength the resulting flux through the read head coil can be found by summing the flux contribution from all these currents.

Consider the arrangement in Fig. 4.10 in which two current-carrying coils are linked by the mutual inductance L_m. The first of the coils generating ϕ_1 flux can be identified with the winding of the read head. The second replaces the magnetization of an elementary volume of the recorded medium by an equivalent ni_2 magnetomotive force which in turn produces the equivalent flux ϕ_2 which would be observed at this point in the medium. Hence,

$$\phi_2 = L_m i_1. \tag{4.23a}$$

$$\phi_1 = L_m i_2. \tag{4.23b}$$

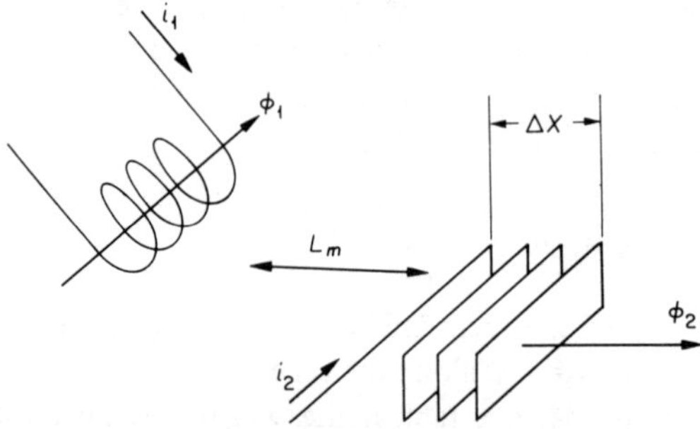

Fig. 4.10. Application of the reciprocity principle in analysis of the reproduction process.

At the low flux levels emanating from the tape the head can be considered as a linear element and as each section of the medium is magnetized to saturation, the medium permeability equals unity. Hence

$$\phi_2 = B_2 A = \mu H_2 A = L_m i_1, \tag{4.24}$$

where A is the cross-sectional area of the volume element in the direction of ϕ_2 and H is the field strength produced at the volume element by the head current. This H is already known from various approximations such as the Karlquist equation; therefore, for a fixed value of y, let

$$H_2 \equiv H_i(x)i_1 \tag{4.25}$$

in which $H_i(x)$ is the field strength per unit current produced by the head. From Equations (4.24) and (4.25)

$$L_m = \mu A H_i(x). \tag{4.26}$$

The next step is to calculate the i_2 equivalent current which is necessary to produce a certain flux density at the volume element. Representing i_2 by a current sheet i_2', the equivalent current per unit distance required to produce the magnetization M within the actual volume element is

$$i_2' = \frac{4\pi}{10} H = \frac{4\pi}{10\mu} M(x) \tag{4.27}$$

The contribution of the incremental flux in a volume element of the length Δx to the read head flux $\Delta\phi$, will be, with L_m from Equation (4.26) and i_2' from Equation (4.27)

$$\Delta\phi_1 = L_m i_2' \, \Delta x = \frac{4\pi}{10} A H_i(x) M(x) \Delta X. \tag{4.28}$$

Hence the total ϕ_1 flux is obtained by

$$\phi_1 = \frac{4\pi}{10} A \int_{-\infty}^{\infty} H_i(x) M(x) \, \mathrm{d}x. \tag{4.29}$$

The reference coordinates for H and M are chosen to account for the relative motion between head and medium as illustrated in Fig. 4.11. Letting $H_i(x) = H(x)$ be the normalized field, the readback voltage, which is proportional to the derivative of Equation (4.29) takes the general form of

$$e(t) = \frac{\mathrm{d}}{\mathrm{d}t} K \int_{-\infty}^{\infty} H(\bar{x}) M(x - \bar{x}) \, \mathrm{d}\bar{x} \tag{4.30}$$

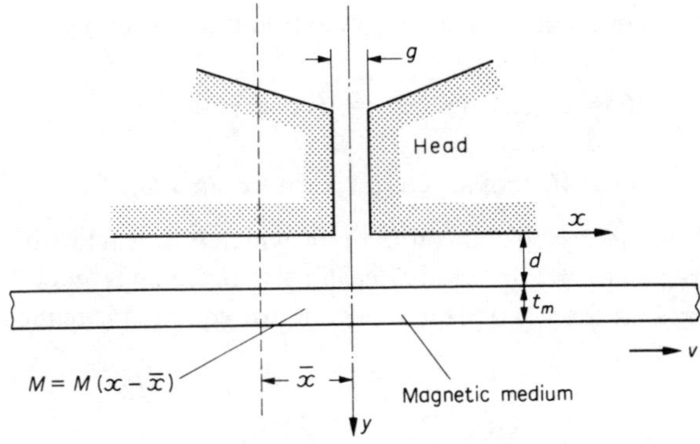

Fig. 4.11. Coordinate system.

Here $\bar{x} = vt$; M is the longitudinal component of magnetization an elementary volume of the medium and H the longitudinal component of the record head field generated by a unit magnetomotive force (i.e. H is defined for $ni = 1$ where n is the number of turns of the write head coil). With surface motion the magnetization pattern is moving relative to the head:

$$M = M(x - \bar{x}).$$

A more appropriate model for magnetic tape will take the H_y field into consideration but it is justified to consider $M(z)$ and $H(z)$ constant. In the y direction the limits are constituted by the medium thickness, hence a more general form of Equation (4.29) is

$$\phi_x = K \int_{t_m}^{t_m+d} \int_{-\infty}^{\infty} M_x(x - \bar{x}, y) H_x(x, y) \, \mathrm{d}x \, \mathrm{d}y. \tag{4.31}$$

Here d is the head-to-medium spacing, t_m the medium thickness and K a constant, and

$$e_x(\bar{x}) = nv \frac{\mathrm{d}\phi}{\mathrm{d}x} = K \int_{t_m}^{t_m+d} \mathrm{d}y \int_{-\infty}^{+\infty} \frac{\partial}{\partial x} M_x(x - \bar{x}, y) H_x(x, y) \, \mathrm{d}x. \tag{4.32}$$

The convolution integral, which appears in various forms in Equations (4.30) and (4.32) links the reproduction voltage with the x component of the recording head field and the medium magnetization and gives a powerful tool for investigation of the reproduction process. Although our concern is digital recording, we will illustrate the use of reciprocity theorem for the very important case of non-saturation sinusoidal magnetization, i.e. analogue recording.

4.10. SINUSOIDAL MAGNETIZATION

Let us consider sinusoidal surface magnetization first, assuming that

$$M_x = M_r \cos k(x - \bar{x}) = M_r \cos \frac{2\pi}{\lambda} (x - vt)$$

$$= M_r (\cos kx \cos \omega t + \sin kx \sin \omega t), \tag{4.33}$$

where $\lambda = v/f$ is the wavelength on the tape which is moving with v velocity and on which f frequency is impressed. Recalling that arctan is an odd function and arctan $(\infty) = \pi/2$, in the $y = 0$ plane the Karlquist equation is reduced to

$$H_x = \begin{cases} 0 & x < -g/2 \\ 1/g & -g/2 < x < g/2 \\ 0 & x > g/2 \end{cases}$$

which is of course the field along the pole faces. The case of particular interest is the thin medium with very small head-to-medium spacing, i.e. $t_m \to 0$, $d \to 0$. Substituting Equation (4.33) into (4.31) we obtain

$$\phi_x = K \int_{t_m}^{t_m+d} \int_{-\infty}^{+\infty} M_r \cos k(x - \bar{x}) H_x(x, y)\, \mathrm{d}x$$

$$= \frac{K}{g} \int_{t_m}^{t_m+d} \mathrm{d}y \int_{-g/2}^{+g/2} M_r \cos kx\, \mathrm{d}x \, (\cos \omega t),$$

or

$$\phi = M_r K t_m \frac{\sin \dfrac{\pi g}{\lambda}}{\pi g/\lambda} \cos \omega t. \tag{4.34}$$

Another case for which the integration can be carried out is that of the remote region when $y > g$.

For this we obtain [1, 10]

$$\phi = K_1 \left(\frac{1 - e^{\frac{-2\pi t_m}{\lambda}}}{2\pi \dfrac{t_m}{\lambda}} \right) \left(e^{-2\pi \frac{d}{\lambda}} \right) \cos \omega t. \tag{4.35}$$

The Equations (4.34) and (4.35) were obtained by W. K. Westmijze [2] using the reciprocity theorem and by R. L. Wallace [1] by the method images. They are of fundamental importance for analogue recording and have been extensively discussed in the relevant literature. Here we only want to draw attention to the interpretation of the above equations.

Equation (4.34) gives the gap loss function as

$$\frac{\sin \dfrac{\pi g}{\lambda}}{\dfrac{\pi g}{\lambda}}, \tag{4.36}$$

i.e. predicts that for $g/\lambda = 1, 2 \ldots n$ the attenuation approaches infinity and the head output will be near-zero. A correction factor introduced by G. J. Fan [11] in its simplest form reduces to

$$\frac{5 - 4\left(\dfrac{\lambda}{g}\right)^2}{4 - 4\left(\dfrac{\lambda}{g}\right)^2}. \tag{4.37}$$

The experimentally observed location of the first zero is near $g/\lambda = 0.88$, in agreement with Fan's correction. In equation (4.35) the factor

$$\frac{1 - e^{-2\pi \frac{t_m}{\lambda}}}{2\pi \frac{t_m}{\lambda}}, \tag{4.38}$$

represents the thickness loss which takes into consideration the finite thickness of the medium and

$$e^{-2\pi(d/\lambda)} \tag{4.39}$$

is the loss caused by the finite head to medium separation. The thickness loss expresses the fact that, particularly at short wavelengths, most of the contribution of the flux through the head comes from the media layers near to the head surface.

The spacing loss is often expressed in decibels and is quoted in the literature as the 'Wallace formula':

$$\text{Spacing loss} = 54.6 \frac{d}{\lambda} \text{ dB},$$

i.e. the head-to-tape separation equal to the wavelength would cause 54.6 dB loss in amplitude of the reproduce head output voltage. D. E. Speliotis et al. [31] have experimentally investigated the head-to-tape separation as the function of bit density in saturation recording for three different read head gap lengths (1000, 150 and 50 μ in). The agreement with the Wallace formula was found to be poor with the exception of the highest densities; however Equation (4.39) can be considered a useful guide for estimating the attenuation caused by head-to-tape separation at high densities.

4.11. STEP FUNCTION MAGNETIZATION

The waveform of the voltage pulse produced for a single, isolated step function change of magnetization is of particular interest, since, as already mentioned earlier, any pattern of magnetization changes between two states (saturation-to-saturation or saturation-to-zero magnetization) can be resolved into a series of step functions. The convolution integral can be easily evaluated for the case when $M_y = 0$, and M_x changes from $-M_r$ to $+M_r$, i.e. the step-function saturation-to-saturation magnetization. For this case we find ([3, 10])

$$e(\bar{x}) = K' \int_{d}^{d+t_m} H_x(\bar{x}, y)\, dy. \tag{4.40}$$

where K' is a constant. The significance of Equation (4.40) is that it relates the reproduce pulse waveform to the fringing field from the recording head, thickness of the medium and head-to-medium spacing. The head field can be computed from the Karlquist equation, or any other convenient approximation.

Equation (4.40) is still not easy to handle; we are looking for special situations when the Karlquist equation takes a simpler form.

An approximation for the field remote from the gap can be written as

$$H' = \frac{1}{\pi r},\tag{4.41}$$

and

$$H'_x = H' \frac{y_n}{r} = \frac{y_n}{\pi r^2} = \frac{1}{\pi} \frac{y_n}{x_n{}^2 + y_n{}^2},\tag{4.42}$$

and therefore

$$e(\bar{x}) = K \int_{d}^{d+t_m} \frac{y}{\bar{x}^2 + y^2}\, dy$$

$$= K \ln \left(\frac{\bar{x}^2 + (d + t_m)^2}{\bar{x}^2 + d^2} \right).\tag{4.43}$$

The maximum value of the pulse will be at $\bar{x} = 0$

$$e_{\max} = e(0) = 2K \ln \frac{d + t_m}{d}.\tag{4.44}$$

Equations (4.43) and (4.44) give a useful indication of the effects of medium thickness and head-to-medium spacing on the reproduce waveform. Consider a constant medium thickness t_m but increasing head-to-tape spacing. As the spacing increases, the amplitude will reduce and the width of the pulse increases; the waveforms will be identical with the x component of the field as shown in Fig. 3.11. Clearly, for a given medium thickness maximum pulse density – as far as the reproduction process is concerned – will be obtained at minimum head-to-medium spacing. Also, as this has been intuitively felt, smallest separation gives highest reproduce voltage. Analysis of the recording process in Section 4.5 indicated that high bit density demands thin recording layers; however reduction in thickness of the recording layer will only be effective if the separation too is reduced. The pulse width at half of the maximum pulse amplitude can be calculated from Equations (4.43) and (4.44) as

$$P_{1/2} = 2\sqrt{(d + t_m)d}.\tag{4.45}$$

The computation can now be extended from the step-function magnetization to other magnetization distribution functions. Using the

$$M_x = \frac{2}{\pi} M_r \tan^{-1}\left(\frac{x}{a} \right),$$

transition function, D. E. Speliotis and J. R. Morrison [9] calculated the output voltage as

$$e(\bar{x}) = K \ln \frac{\bar{x}^2 + (d + t_m + a)^2}{\bar{x}^2 + (d + a)^2},\tag{4.46}$$

and the half-pulse width

$$P_{1/2} = 2\sqrt{(d + t_m + a)(d + a)}. \tag{4.47}$$

The comparison of Equation (4.47) with (4.45), which was derived for the case of step-function transition, shows the effect of the spreading of the transition zone on the amplitude and width of the readback pulse. For thin surfaces $d + a \gg t_m$ and the binomial expansion of Equation (4.47) gives

$$P_{1/2} \simeq 2(a + d) + t_m. \tag{4.48}$$

Substituting $a = 2t_m(M_r/H_c)$ from Equation (4.17) we obtain an approximation for the half-pulse width as the function of the medium magnetic parameters, thickness and head-to-medium separation as

$$P_{1/2} \simeq t_m \left(1 + 4\frac{M_r}{H_c}\right) + 2d. \tag{4.49}$$

The general case of finite reproduce head gap length and an arctan magnetization transition leads to a lengthy and somewhat cumbersome expression for the reproduce voltage and half-pulse width as the function of g gap length, t_m medium thickness, d separation and a demagnetization parameter. However, for the case of thin media for which $t_m \ll d + a$ it takes the more manageable form of [9, 12, 24]

$$e(\bar{x}) = \frac{K}{g} \left(\tan^{-1} \frac{\bar{x} + g/2}{d + a} - \tan^{-1} \frac{x - g/2}{d + a}\right), \tag{4.50}$$

and the peak reproduce voltage will be

$$e_0 = 2\frac{K}{g} \tan^{-1} \frac{g/2}{a + d}. \tag{4.51}$$

The constant K includes the medium parameter M_r, tape speed v, number of turns n, track width w, medium thickness t_m, head efficiency factor η_K and numerical constant k_1

$$K = k_1 4vnwM_r t_m \eta_R \tag{4.52}$$

The half pulse-width can again be computed by solving the $e(\bar{x}) = e_0/2$ equation for $\bar{x}$ [24]

$$P_{1/2} = 2\sqrt{(a + d)^2 + (g/2)^2}. \tag{4.53}$$

Many slightly different approximations are quoted in the literature for the width and amplitude of a pulse generated by the reproduction of an isolated transition [12, 24, 27, 45, 48, 60, 61]. The semi-empirical relation of R. O. McCary [12] gives the half-pulse width as

$$P_{1/2} = \sqrt{g^2 + l^2 + 4d(d + t_m)}, \tag{4.54}$$

and the estimate for l, the length of the transition region after demagnetization is

$$l = 0.5\, t_m \left(\frac{Br}{H_c} - 1 \right). \qquad (4.55)$$

An extensive study of the dependence of half-pulse width on magnetic parameters, assuming a constant gap length and head-to-medium separation was carried out by D. E. Speliotis [27] who concluded that the relation

$$P_{1/2}(H_c, B_r, t_m) = CH^{\alpha}B^{\beta}t_m{}^{\gamma} \qquad (4.56)$$

holds over a wide range of coercivities and remanences. Here C is a constant. For 'thick' material it is only the constant γ which differs substantially from the value derived for the thin material, the α and β exponents remain nearly constant. The constant α is negative. On the other hand, J. C. Mallinson's [39] analysis of the theoretical limit to digital pulse resolution does not support the assumption that the relation between pulse width and tape parameters can be expressed in the form of power law for a wide range of parameter variation. The conclusion of the numerical results in [39] is that the output voltage and the pulse width increase with increasing medium thickness, but the effects of coercivity, remanence, and squareness ratio do not follow a simple pattern.

The recording head gap length does not enter explicitly into the equations derived for the reproduce pulse width, whereas the reproduce head gap length is one of the basic parameters which determine the head performance. It has often been stated that the record head gap length has a second-order effect only, and a wider gap has an advantage over a narrow because it permits the saturation of the tape with a low head current. However, the recording head gap length should be in proportion to the medium thickness, and narrow recording head gaps give the highest densities on thin films.

4.12. PULSE ASYMMETRY

The readback pulse from an isolated transition shows a characteristic asymmetry: the decay at the trailing edge is more gradual than at the rising slope. The asymmetry has been attributed to the y component of magnetization, [3, 10 (p. 68 and 85), 15], to the asymmetry of the transition zone [33, 56], to read/write electronics and time lags caused by head inductance and eddy current effects [62]. The y component of the recording field, as was illustrated in Fig. 3.12, reverses polarity at the gap centre line. As the magnetizing field will be the vector sum of the fields around or past the trailing edge of the gap, it is an obvious assumption that any significant degree of magnetization in the y direction will cause distortion of the readback pulse because the nature of the y component contribution to the output signal. Numerical computations of the readback pulse waveform, taking into account the y component of magnetization, have been carried out by Potter [15] for some M_y/M_x ratios, and the corresponding plots produce the distortion similar to the actual waveforms. However, it is somewhat difficult to explain the observed pulse asymmetry with increasing recording current; this requires the assumption of changing M_y/M_x ratios.

The shape of the transition zone computed with the iterative process shows a marked asymmetry, even if only that component of magnetization is considered which is in the direction of motion [56]. The asymmetry of the readback pulse can be attributed to the asymmetrical writing by the head field away from the gap centre line. As the recording current increases, the transition region moves further away from the centre line of the recording head and becomes more asymmetrical. This can explain the dependence of the observed asymmetry on the write current. Self-demagnetization plays an important role. If self-demagnetization and hence the transition zone is small, the output waveform is highly affected by the losses of the reproduction process. The reproduction losses tend to mask the asymmetry and a nearly symmetrical pulse is obtained. On the other hand if self-demagnetization losses are large, and consequently transition zone length is increased, the self-demagnetization itself tends to make the recorded magnetization symmetrical. Hence, the asymmetry of the readback pulse, as observed at the readback amplifier terminals, will be a complex function of a number of significant parameters of the system: recording current, medium magnetic parameters and thickness, reproduce head losses. At high recording densities the asymmetry of a single-transition readback pulse is of little practical significance; the effects of adjacent pulse transitions are far more important and tend to mask the isolated pulse asymmetry.

4.13. PEAK SHIFT AND PULSE CROWDING

In the previous paragraphs where the digital recording and reproduction process was analysed we always assumed an 'isolated' transition, i.e. the effect of adjacent pulses was considered negligible. Any recorded pattern can be considered as a series of step function magnetization changes but, with closely spaced transitions, the interaction between transitions must be taken into consideration.

The external field generated by a recorded transition is very low thus the reproduce head can be considered as a linear element and the principle of superposition can be applied [10 (p. 117), 49, 50, 52, 63]. The net voltage, at the reproduce head terminals, which is caused by a series of step-function changes in the medium, is obtained by linear superposition of the individual voltages. This is illustrated in Fig. 4.12 with a positive/negative saturation-to-saturation sequence of transitions. As long as the adjacent transitions are far from each other, there is no interaction (Fig. 4.12(a)). As the transitions are moved closer to each other (Fig. 4.12(b)) the interaction between the two pulses causes a change in amplitude and the separation between the two peaks s, is greater than the separation h between the corresponding write current reversals. In Fig. 4.12(b) $+e(x)$ and $-e(\bar{x})$ represent the output voltage from the positive and negative going transition, and $e_r(\bar{x})$ the resultant obtained by superposition. A superposition of a train of closely spaced transitions will show that the interaction causes a reduction of pulse amplitude, a shift in the location of the peak and a shift in the base line of the resultant signal.

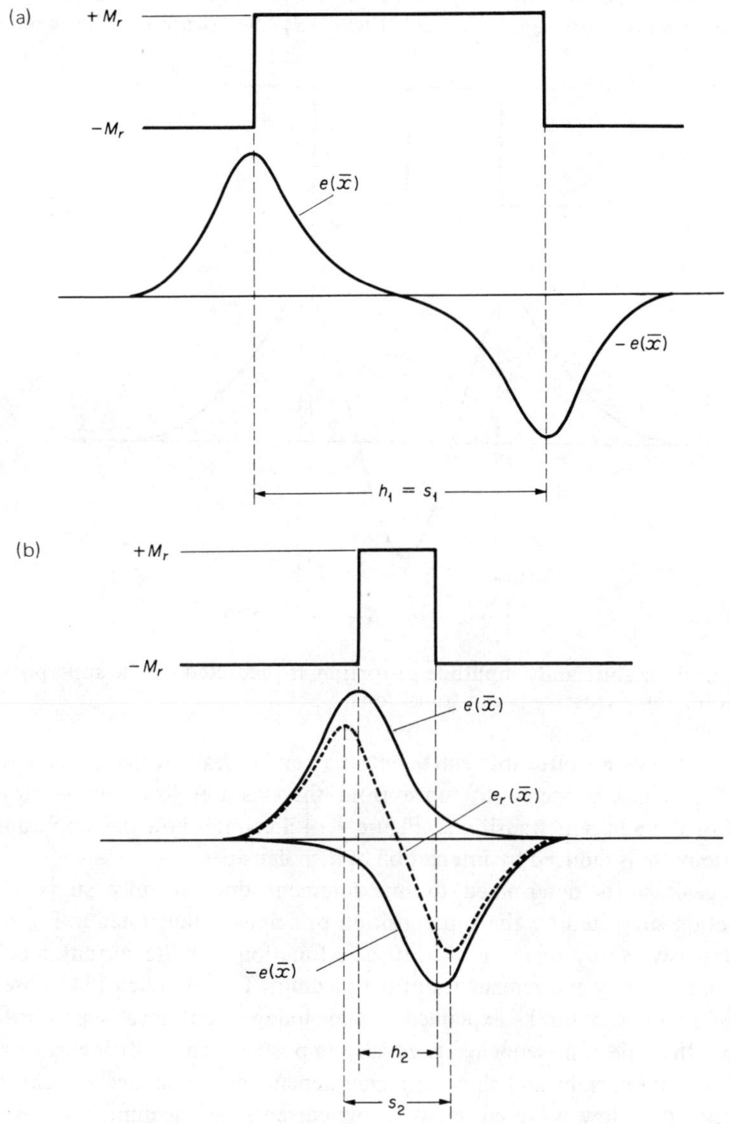

Fig. 4.12. Two adjacent saturation reversals.

The application of the superposition principle in determination of the resultant amplitude and peak shift seems to be justified as long as the density of saturation reversals does not approach the transition length. When the separation between reversals is less than the transition length, the transition region will be modified by the subsequent magnetization change. At this point the nonlinear nature of the writing process invalidates the basic assumptions made when applying the superposition principle.

In the same manner as in Fig. 4.12, the output signal from three adjacent saturation reversals can be constructed (Fig. 4.13); the position of the two outer

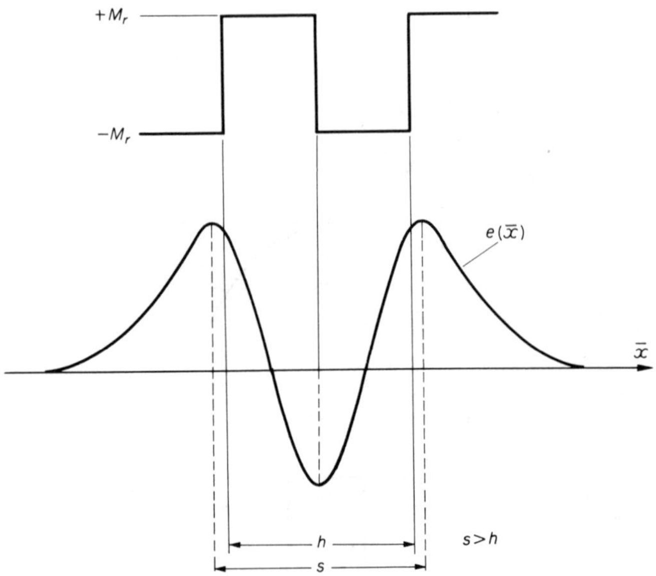

Fig. 4.13. Peak shift and amplitude distortion as predicted by the superposition theorem for three closely-spaced transitions.

transitions shows an outwards shift, but the centre peak, which is reduced in amplitude, is not expected to move from the position it would occupy in absence of the adjacent transitions. Figure 4.14 illustrates how the amplitude of the centre pulse is reduced by interaction at high densities.

The peak shifts determined by measurement do not fully support the construction suggested by the superposition principle as illustrated in Fig. 4.12. In his extensive study of the peak shift as a function of write current, medium thickness, coercivity and remanence of the medium, T. J. Beaulieu [51] observes peak shifts which cannot be explained by time-independent linear superposition; it appears that the time-sequence is equally important. The shift increases with increasing write current and shows a higher dependence on medium thickness at high rather than low write currents; at low currents the medium thickness has little effect on the peak shift.

The magnetic parameters needed to reduce the peak shift were shown to depend on the relative position of the peak in the sequence, the medium thickness and write current level. The medium, with maximum H_c/M_r ratio, does not exhibit the minimum peak shift. In all measurements the record and reproduce head gap lengths were kept constant, thus the effect of gap length did not appear in the results.

It has been observed that for two adjacent transitions as shown in Fig. 4.12, for which the superposition theorem predicts symmetrical shifts, the actual peak shifts are different for the positive and negative transitions. This is explained by S. Iwasaki and T. Suzuki [43] as the effect of the demagnetizing field. A study of peak shift by B. Kostyshin [55] confirms that the position of the peak of the recorded transition depends on the polarity of the DC bias; however B. Kostyshin concludes that, by appropriate choice of the system design parameters, gap length of the read/write head, recording medium parameters, and

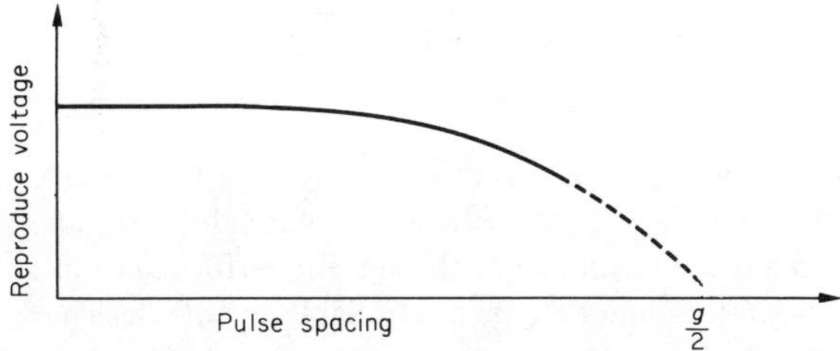

Fig. 4.14. Output voltage versus recording density (pulse crowding).

write current rise time, the peak shift can be virtually eliminated. The coatings used on tape permit only small variations of the magnetic parameters, hence the conclusions are applicable to thin film disc coatings.

Figure 4.13 illustrates the reduction in the amplitude of the reproduced pulse as the function of recorded transition density. As long as the minimum spacing between two saturation reversals is large and there is no interaction between pulses, the output voltage peak amplitude will remain unchanged. When the interference of adjacent pulses extends to the centre of the pulse, the amplitude is reduced as the adjacent pulse is of opposite polarity. As the width of the reproduced pulse – assuming that a and d are small in Equation (4.53) with respect of the reproduce head gap length g – is primarily determined by the reproduce head gap length, a case which is true for many actual tape systems; the reproduce head output will fall to zero at a spacing near $g/2$ (Fig. 4.14).

4.14. SYSTEM OPTIMIZATION

The analysis of digital recording and reproduction in the previous sections has led, in spite of the complex processes involved, to some remarkably simple approximations for the amplitude and width of the readback voltage, for an isolated transition between two oppositely magnetized states. For thin mediums, the readback pulse voltage was approximated by

$$e\left(\bar{x}\right) = \frac{K}{g}\left(\tan^{-1}\frac{\bar{x}+g/2}{d+a} - \tan^{-1}\frac{x-g/2}{d+a}\right), \qquad (4.57)$$

where $K = k_1 4vnwM_r t_m \eta$, and k_1 is a numerical constant depending on the system of units employed. The peak value of the readback pulse from Equation (4.51) is:

$$e_0 = 2\frac{K}{g}\tan^{-1}\frac{g/2}{d+a}. \qquad (4.58)$$

A good approximation for the half-pulse width is: $(t_m \ll d+a)$

$$P_{1/2} = 2\sqrt{(a+d)^2 + (g/2)^2}. \qquad (4.59)$$

Equations (4.57) through (4.59) link the system parameters – gap length, head-to-medium separation, head efficiency – with the medium parameters M_r, H_c, and t_m, and allow an estimate of the system performance in terms of factors which are, at least to some extent, controllable by the designer. A convenient measure of the system performance is the ratio of peak reproduce voltage to half pulse width, $e_0/P_{1/2}$; this can be maximized with respect to the system parameters.

Let us now review the messages contained in Equation (4.57) through (4.59); some of them are quite obvious, others open to interpretation. The head-to-medium separation should be, quite clearly, as small as possible. We have encountered two limiting factors such as head and medium surface roughness, and the air film which builds up between head and tape at higher tape speeds, even in 'in-contact' recording. It is also obvious that the factor a should be small which implies thin medium with high coercivity. Reducing the medium thickness causes loss of amplitude, hence M_r must be increased at the same time so that $M_r t_m$ remains nearly constant. The reproduce gap length should be related to medium thickness and the M_r/H_c ratio; no substantial improvement in resolution will be obtained by reducing the reproduce gap length if a and d are the dominant factors in Equation (4.59). The permeability of the medium, and the length of the write head gap are of secondary importance. The signal-to-noise ratio does not appear explicitly but as the signal amplitude is – as a first approximation – linearly dependent on the medium thickness, an 'optimum' medium thickness different from zero can be defined yielding the lowest acceptable signal-to-noise ratio. The reproduce head parameters enter into Equation (4.58) via the efficiency factor η_R which is in turn a function of core and gap reluctance ratios; again an optimum gap length may be calculated but technological consideration may not permit to implement the design.

REFERENCES

1. Wallace, R. L. Jun. (1951), 'The reproduction of magnetically recorded signals', *Bell Syst. Tech. J.*, **30**, 1146–1173.
2. Westmijze, W. K. (1953), 'Studies on magnetic recording', *Philips Research Reports*, **8**, 245–269.
3. Eldridge, D. F. (1960), 'Magnetic recording and reproduction of pulses', *IRE Trans. Audio*, **Au-8**, 47–52.
4. Teer, K. (1961), 'Investigation of the magnetic recording process with step functions', *Philips Research Reports*, **16**, 469–491.
5. Barkouki, M. F. and Stein, I. (1963), 'Theoretical and experimental evaluation of RZ and NRZ record characteristics', *IEEE Trans. Electronic Computers*, **EC-12**, 92–100.
6. Stein, I. (1963), 'Generalized pulse recording', *IEEE Trans. Electronic Computers*, **EC-12**, 77–92.
7. Miyata, J. J. and Hartel, R. R. (1959), 'The recording and reproduction of signals on magnetic medium using saturation-type recording', *IRE Trans. Electronic Computers*, **EC-8**, 159–169.
8. Aharoni, A. (1966), 'Theory of NRZ recording', *IEEE Trans. Mag.*, **Mag-2**, 100–109.
9. Speliotis, D. E. and Morrison, J. R. (1966), 'A theoretical analysis of saturation magnetic recording', *IBM J. Res. and Dev.*, **10**, 233–243.
10. Hoagland, A. S. (1963), 'Theory of the digital magnetic recording process', *Digital Magnetic Recording*, Wiley, 56–78.
11. Fan, G. J. (1961), 'A study of the playback process of a magnetic ring head', *IBM J. Res. and Dev.* 321–325.
12. McCary, R. O. (1971). 'Saturation magnetic recording process', *IEEE Trans. Mag.*, **Mag-7**, 1, 4–16.
13. Feth, G. O. (1962), 'Analysis of magnetic recording fields', *AIEE Trans. (Commun. Electronics)*, **81**, 267–279.
14. Tjaden, D. L. A. (1964), 'The magnetic recording process—looked at through a magnifying glass', *Proc. Int. Conf. Magnetic Recording, IEEE, London*, 135–138.
15. Potter, R. I. (1970), 'Analysis of saturation magnetic recording based on arctangent magnetization transitions', *J. Appl. Phys.*, **41**, 1647–1651.
16. Tjaden, D. L. A. (1963–64), 'A 5000 : 1 scale model of the magnetic recording process', *Philips Tech. Rev.*, **25**, 319–329.
17. Curland, N. and Speliotis, D. E. (1970), 'A theoretical study of an isolated transition using an iterative hysteretic model', *IEEE Trans. Magn.*, **Mag-6**, 3, 640–646.
18. Speliotis, D. E. and Judy, J. H. (1970), 'Calculations of external bit fields', *IEEE Trans. Mag.*, **Mag-6**, 3, 653.
19. Davies, A. V., Middleton, B. K. and Tickle, A. C. (1965), 'Digital recording properties of evaporated cobalt films', *IEEE Trans. Mag.*, **Mag-1**, 344–348.
20. Nishikawa, M. (1968), 'Digital recording properties of relatively thick magnetic medium', *IEEE Trans. Mag.*, **Mag-4**, 286–290.

21. Morrison, J. R. (1968), 'A study of the effect of remanence and thickness on the recording properties of thick particulate media', *IEEE Trans. Mag.*, **Mag-4**, 281–286.
22. Curland, N. and Speliotis, D. E. (1970), 'Transition region in recorded magnetization patterns', *J. Appl. Phys.*, **41**, 3, 1099–1101.
23. Della Torre, E. (1960), 'The influence of magnetic tape on the field of a recording head', *RCA Review*, **21**, 45–51.
24. Bonyhard, P. I., Davies, A. V. and Middleton, B. K. (1966), 'A theory of digital magnetic recording on metallic films', *IEEE Trans. on Mag.*, **Mag-2**, 1, 1–5.
25. Francis, E. E. and Ku, T. C. (1962–63), 'A theoretical solution for the magnetic field in the vicinity of a recording head air gap', *IBM J. Res. and Dev.*, **6**, 260–262; *IBM J. Res. and Dev.*, 7, 4, 355.
26. Duinker, S. (1957), 'On the resolving power in the process of magnetic recording', *Tijdschrift van het Nederlands Radiogenootschap.*, **22**, 29–48.
27. Speliotis, D. E. (1969), 'The effect of remanence in digital recording', *IEEE Trans. Mag.*, **Mag-5**, 3, 253–258.
28. Speliotis, D. E. (1968), 'Theory and experiment in magnetic recording', *J. Appl. Phys.*, **39**, 1310–1317.
29. Speliotis, D. E. (1967), 'Magnetic recording theories: accomplishments and unresolved problems', *IEEE Trans. Mag.*, **Mag-3**, 195–200.
30. Uhl, K. (1966), 'Contribution to recording and reproduction with NRZ method', *IEEE Trans. Mag.*, **Mag-2**, 3, 221–224.
31. Speliotis, D. E., Bate, G., Morrison, J. R. and Braun, R. E. (1965), 'An investigation of separation losses in high-speed, high density recording tapes', *IEEE Trans. Mag.*, **Mag-1**, 101–104.
32. Schools, R. S. (1961), 'Recording and reproduction of NRZ1 signals', *J. Appl. Phys.*, **32**, 425–435.
33. Morrison, J. R. (1969), 'An analysis of recording demagnetization', *IEEE Trans. Mag.*, **Mag-5**, 4, 948–954.
34. Herbert, D. J. and Patterson, D. W. (1966), 'A computer simulation of the magnetic recording process', *IEEE Trans. Mag.*, **Mag-1**, 236–242.
35. Mee, C. D. (1964), *The physics of magnetic recording,* Chapter 4, 85–138, North Holland Publishing Co.
36. Karlquist, O. (1954), 'Calculation of the magnetic field in the ferromagnetic layer of a magnetic drum', *Trans. Royal Inst. of Tech.*, Stockholm, **86**, 3–27.
37. Mallinson, J. C. (1966), 'Demagnetization theory for longitudinal recording', *IEEE Trans. Mag.*, **Mag-2**, 233–235.
38. Bayer, R. G. (1963), 'A theoretical treatment of self-demagnetization in magnetic recording', *IEEE Trans. Audio*, **Au-11**, 81–88.
39. Mallinson, J. C. (1969), 'A theoretical limit to the digital pulse resolution in saturation recording', *IEEE Trans. Mag.*, **Mag-5**, 2, 91–97.
40. Schwantke, G. (1957), 'Beitrag zur Darstellung des Spaltfeldes beim Magnetton', *Acustica*, 7, 363–371.

41. Westmijze, W. K. (1952), 'Gap-length formula in magnetic recording', *Acustica*, **2**,

42. Kostyshin, B. (1966), 'A theoretical model for a quantitative evaluation of a magnetic recording system', *IEEE Trans. Mag.*, **Mag-2**, 236–242.

43. Iwasaki, S. and Suzuki, T. (1968), 'Dynamical interpretation of magnetic recording process', *IEEE Trans. Mag.*, **Mag-4**, 3, 269–276.

44. Kostyshin, B. (1962), 'A harmonic analysis of saturation recording in a magnetic medium', *IRE Trans. Electronic. Computers*, **EC-11**, 253–256.

45. Middleton, B. K. (1966), 'The dependence of recording characteristics of thin metal tapes on their magnetic properties and on the replay head', *IEEE trans. Mag.*, **Mag-2**, 3, 225–229.

46. Speliotis, D. S. and Morrison, J. R. (1966), 'Correlation between magnetic and recording properties in thin-surfaces', *IEEE Trans. Mag.*, **Mag-2**, 3, 208–212.

47. Herbert, J. R. (1966), 'The readback process in digital magnetic recording', *IEEE Trans. Mag.*, **Mag-2**, 247–250.

48. Mallinson, J. C. and Steele, C. W. (1971), 'A computer simulation of unbiased sine wave recording', *IEEE Trans. Mag.*, **Mag-7**, 2, 249–254.

49. Hoagland, A. S. and Bacon, G. C. (1960), 'High-density digital magnetic recording techniques', *IRE Trans. Electronic Computers*, **EC-9**, 2–11.

50. Templeton, H. S. and Bate, G. (1963–64), 'Peak shift study in high-density magnetic tape recording', IEEE International Conference on Nonlinear Magnetics, Washington, D.C., *IEEE Trans. Comm. and Electr.*, **83**, 429–432.

51. Beaulieu, T. J. (1969), 'The contribution of the magnetic medium to phase shift and resolution in magnetic recording', *IEEE Trans. Mag.*, **Mag-5**, 3.

52. Morrison, J. R. and Speliotis, D. E. (1967), 'Study of peak-shift in thin recording surfaces', *IEEE Trans. Mag.*, **Mag-3**, 208–211.

53. Dingwall, P. F. and Labrum, G. R. (1965), 'High density digital recording on magnetic tape', *Royal Aircraft Establishment Technical Report No. 65006*.

54. Potter, R. I. and Schmulian, R. J. (1971), 'Self-consistently computed magnetization patterns in thin film magnetic media', *IEEE Trans. Mag.*, **Mag-7**, 4, 873–880.

55. Kostyshin, B. (1971), 'The write process in magnetic recording', *IEEE Trans. Mag.*, **Mag-7**, 4, 880–885.

56. Curland, N. and Speliotis, D. E. (1971), 'An interative hysteretic model for digital magnetic recording', *IEEE Trans. Mag.*, **Mag-7**, 3, 538–543.

57. Chapman, D. W. (1963), 'Optimizing the digital magnetic recording process', *Proc. IEEE*, **51**, 247–248.

58. Ku, T. C. (1961), 'An analytical expression for describing the write process in magnetic recording', *Proc. IRE*, **49**, 8, 1337–1338.

59. Geurst, J. A. (1963), 'The reciprocity principle in the theory of magnetic recording', *Proc. IEEE*, **51**, 11, 1573–1577.

60. Chao, S. C. (1965), 'On the output pulse width of digital recording', *Proc. IEEE*, **53**, 2, 193–194.

61. Sierra, H. H. (1965), 'Estimate of pulse width in digital magnetic recording', *Proc. IEEE*, **53**, 5, 513–514.

62. Davies, A. V. (1964), 'The influence of some head and coating properties on pulse resolution in non-return to zero digital recording', *International Conference on Magnetic Recording, London*, 68–71.

63. Mallinson, J. C. and Steel, C. W. (1969), 'Theory of linear superposition', *IEEE Trans. Mag.*, **Mag-5**, 4, 886–890.

64. Siakkou, M. (1969), 'Influence of coating permeability in thin film pulse recording', *IEEE Trans. Mag.*, **Mag-5**, 4, 891–894.

65. Eppstein, E. D. and Morrison, J. R. (1969), 'An analysis of recording demagnetization', *IEEE Trans. Mag.*, **Mag-5**, 3, 188.

66. Chapman, D. W. (1963), 'Theoretical limit on digital magnetic recording density', *Proc IEEE*, **51**, 394–395.

67. Mallinson, J. C. (1967), 'Novel technique for the measurement of demagnetizing fields of longitudinal recording', *IEEE Trans. Mag.*, **Mag-3**, 3, 201.

68. Smaller, P. (1966), 'An experimental study of short wavelength recording phenomena', *IEEE Trans. Mag.*, **Mag-2**, 3, 242–246.

69. George, D. J., King, S. F. and Carr, A. E. (1971), 'A self-consistent calculation of the magnetic transition recorded on a thin-film disc', *IEEE Trans. Mag.*, **Mag-7**, 240–243.

70. Nishikawa, M. (1970), 'Characteristics of readback signal in digital magnetic recording', *IEEE Trans. Mag.*, **Mag-6**, 811–817.

71. Greiner, J. (1956), 'Feldstarke und Spaltverteilungsfunktion eim Sprechkopf mit und ohne Band', *Nachrichtentechnik*, **6**, 63–70.

72. Schwantke, G. (1957), 'Beitrag zur Darstellung des Spaltfeldes beim Magnetton', *Acustica*, **7**, 6, 363–371.

73. Pfirnman, V. K. (1967), 'Method of determining of magnetic field of a recording head by using tapes of various coating thickness', *IEEE Trans. Mag.*, **Mag-3**, 4, 625–627.

74. Francis, E. E. and Ku, T. C. (1962–63), 'A theoretical solution for the magnetic field in vicinity of a recording head airgap', *IBM. J. of Res. and Dev.*, **6**, 2, 260–262; 7, 4, 355.

75. Hawkins, J. K. (1968), *Circuit design of digital computers*, Magnetic-surface recording, 460–474, J. Wiley & Sons.

76. Sauter, G. F., Paul, M. C., Oberg, P. E. and Kaske, A. D. (1972), 'Transverse recording using thin film recording heads', *IEEE Trans. Mag.*, **Mag-8**, 2, 194–200.

77. Fischer, R. D. and Blades, J. D. (1972), 'Recording gap fields by Lorentz shadographs and characteristics of single crystal MnZn ferrites', *IEEE Trans. Mag.*, **Mag-8**, 2, 232–238.

78. Mallinson, J. C. (1972), 'Optimum media thickness for digital recording', *IEEE Trans. Mag.*, **Mag-8**, 2, 239–240.

79. Kosters, A. J. and Speliotis, D. E. (1971), 'Predicting magnetic recording performance by using single pulse superposition', *IEEE Trans. Mag.*, **Mag-7**, 3, 544.

80. Pear, C. B. Jun. (1969), 'Write-current pulse shape in NRZ recording', *IEEE Trans. Mag.*, **Mag-5**, 3, 190–191 (Abstract).
81. Nakamura, Y. and Iwasaki, S. (1969), 'The relationship between the scalar and the vector magnetization in the theory of magnetic recording', *IEEE Trans. Mag.*, **Mag-5**, 190–191 (Abstract).
82. Brown, W. F. (1962), *Magnetostatic principles in ferromagnetism*, North-Holland Publishing Company, 18–28.
83. Jacoby, G. V. (1968), 'Signal equalisation in digital magnetic recording', *IEEE Trans. Mag.*, **Mag-4**, 3, 302–305.

5 Digital recording methods

5.1. INTRODUCTION

Digital magnetic tape storage systems use almost exclusively saturation type recording where either the presence or absence, or alternatively the polarity and not the amplitude of the signal, carries the information. Data to be stored on tape are usually presented as a stream of pulses. The logic of the magnetic tape store converts the pulses into magnetized areas on the tape. This latter process is described under various names such as recording method, recording system, modulation or encoding.

The format of the recording may be either series or parallel. In series format the train of pulses representing the digits of an alphanumeric character (i.e. letters of the alphabet and numbers) are recorded bit-serial and character-serial on one (or sometimes two) tracks. The data are interspersed by additional pulses for error checking, synchronization, marking the beginning and end of a block, record, or file. This series recording allows high linear recording density without excessive demands for close tolerances on heads and transport; on the other hand, it makes poor use of the tape as storage medium and is primarily employed in small temporary or intermediate stores such as cassette tape systems, which will be discussed in Chapter 7.

Most bulk magnetic tape data storage systems use bit-parallel character-serial recording. As in many computers a group of six or eight bits are manipulated and transferred as a unit, the choice of a format on the tape which consists of six or eight data tracks and one further channel reserved for an error-checking (parity) bit is fairly obvious. The format is similar to the one used on punched paper tape, but the bit density is – in the standardized format for information interchange – up to 1600 characters per inch as opposed to the 10 characters/inch on paper tape.

In high density parallel recording systems the tolerances on the heads and transport are extremely tight. The recording system must ensure that the data bits in one frame are not scattered around the ideal centre line of the frame by more than a fraction of the frame to frame spacing. The recording system errors include the static and dynamic skew – the errors caused by the read/write head gap scatter and head mounting tolerances, the imperfections of the tape transport and tape width variations – and the errors caused by electronic parameter tolerances such as variation of impedances between tracks in one head, and variation in write current which causes a spread in the write time and location of the recorded peak. The time-synchronism of bits of one character which are recorded on parallel channels can be improved by electronic skew correction.

A simple de-skewing circuit may contain a variable delay for each track which eliminates the effects of the static skew, i.e. the constant component of the skew caused by read/write head gap scatter. A more complex skew correction system has, in addition to the variable delay static correction, a shift register per track. As the data are read from the tape, each subsequent bit in any channel 'pushes' the bit in front of it farther along the shift register. When any bit reaches the end of the register, this bit and the other bits which belong to the same frame are read and removed. The bits of any one frame always remain identifiable because eventually they become the 'next in line' for removal and large skew errors can be corrected.

5.2. ENCODING AND DECODING

The terms 'code', 'coding' are used in the technical literature in several different senses; in this context 'encoding' is used to describe the method of converting digital information into magnetized areas on the tape, and 'decoding', the recovery of the original data from the tape. The coding system has to provide the maximum bit packing density consistent with reliability, low sensitivity to imperfections of the tape and transport, and simplicity of circuitry. In general, increased density of information can be achieved at the expense of increased error rate and limited exchangeability of recorded tapes between machines.

The reading (decoding) operation is concerned with the reconstruction of the original binary input data from the read output waveform. The time interval within which the decision whether the readback signal is a logical zero or one, is defined by a 'clock' signal. This clock determines the boundaries of a 'cell' to which each bit is confined. The clock may either be generated externally or derived from the data itself when the system is referred to as self-clocking. The external clocking becomes impractical at high bit densities as it requires extreme precision and stability of the tape motion and data logic otherwise the synchronism between clock and data is soon lost.

A large variety of coding systems have been proposed and investigated [1, 2, 3, 4, 7, 14, 25, 29] but few have been implemented and gained wider acceptance. It is debatable whether an 'optimum' code exists for any particular system; the criteria for the choice of a code may be the maximum recording density which is obtainable with fixed magnetic and mechanical parameters as the function of the coding, the reliability of data, and the complexity of encoding and decoding circuits. The optimum code may be different for multitrack parallel recording and single-track serial recording on moving head disc where interaction between heads does not need to be considered.

A list of the most important features of various codes can be drawn up; we will follow here the classification of A. E. Whitehouse [27]. Some features will be rated as 'good', 'fair' or 'poor', others as 'high' or 'low'. The order of entries is rather arbitrary and does not indicate their importance in selecting a code.

1. Efficiency. Code efficiency can be defined as the number of stored bits per flux reversal (in percent).

2. Inter-symbol correlation. In choosing a waveform set to represent different characters, it is desirable to make each waveform as different as possible from all others. For a binary system set, this occurs when symbols are identical but of opposite sign. We will call this correlation as 100%.

3. Bandwidth. A code requiring narrow bandwidth is preferable to one with need for wider bandwidth. Also, a code which necessitates DC, or very low frequency response, is undesirable because it lengthens amplifier settling time, rules out transformer coupling for low-inductance heads, and increases crosstalk at a wavelength on the tape which is comparable to pole-length of the head.

4. Self-clocking. Self-clocking capability is highly desirable, particularly for high-density recording.

5. Read resolution. The code should permit the utilization of the full cell length for decision-making whether the signal represents a logical zero or one (in a two-symbol code).

6. Complexity of encoding and decoding circuits. The cost, per bit, is dependent on the complexity and tolerances of the circuit elements which are required for encoding and decoding.

7. Noise immunity. The sensitivity to drop-outs, drop-ins and inter-track crosstalk varies from code-to-code.

A number of other factors which should be considered in the evaluation of a code, are implicit in the features already listed. Such factors are the ease of error detection, the maximum variation of peak shift and amplitude at a specified bit density for the worst-case sequence of bits, and finally the all-important problem of how the choice of code effects the error rate.

Figure 5.1 shows the tape magnetization for some selected codes which have been extensively used on earlier systems, and others which are currently employed in the overwhelming majority of magnetic tape computer stores. It should be noted that various textbooks and international standards use different terminology for description of the same coding system. The most significant NRZ1 and PE coding methods are discussed in some detail.

1. Return-to-zero (RZ). In this method a binary one is represented by saturating the tape in one direction for a short period; the tape itself, is not magnetized for a binary zero. The width of the pulse representing a binary one is governed by the tape and head characteristics; if too short, the reproduce voltage will be

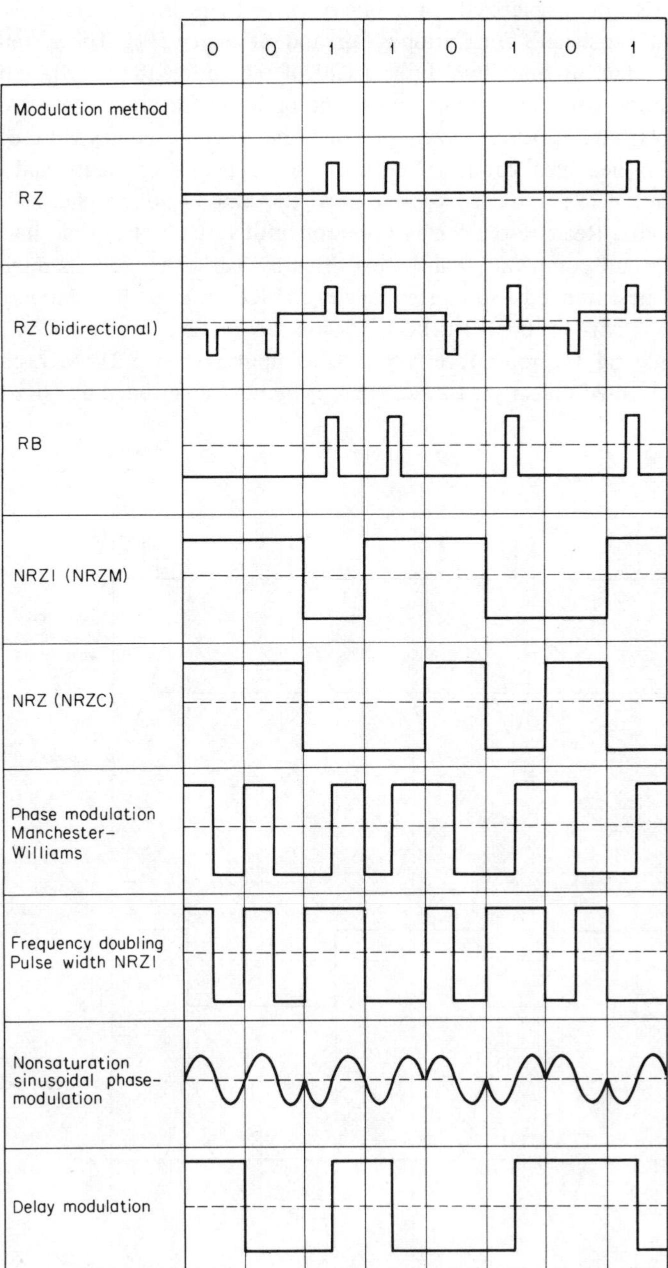

Fig. 5.1. Tape magnetization for various coding methods.

too low, if wider than necessary the maximum bit packing density is reduced. As the tape is not magnetized for a binary zero, there is no difference between 'zero' and 'no signal'. Correlation is fair, and efficiency 50%. The system needs a clock to establish time-slots. In a variant of the above, the tape is saturated in one direction for a binary one and in the opposite for binary zero. As there is now a pulse present for both zero and one, the system becomes self-clocking and correlation becomes good. In both cases the recording signal and the tape magnetization are 'returned to zero', i.e. the tape is not magnetized between binary digits. Read resolution is poor for both variants, less than half of a bit period. Circuit complexity can be classified as 'fair'. The read channel response needs to be extended to low frequencies for RZ, but not for bidirectional RZ. Noise sensitivity is poor for both versions. A bit is represented by two flux transitions and the reproduce voltage is a 'dipulse' (Fig. 5.2). NRZ coding was extensively used on early data stores but is by now superseded by NRZ1 and PE.

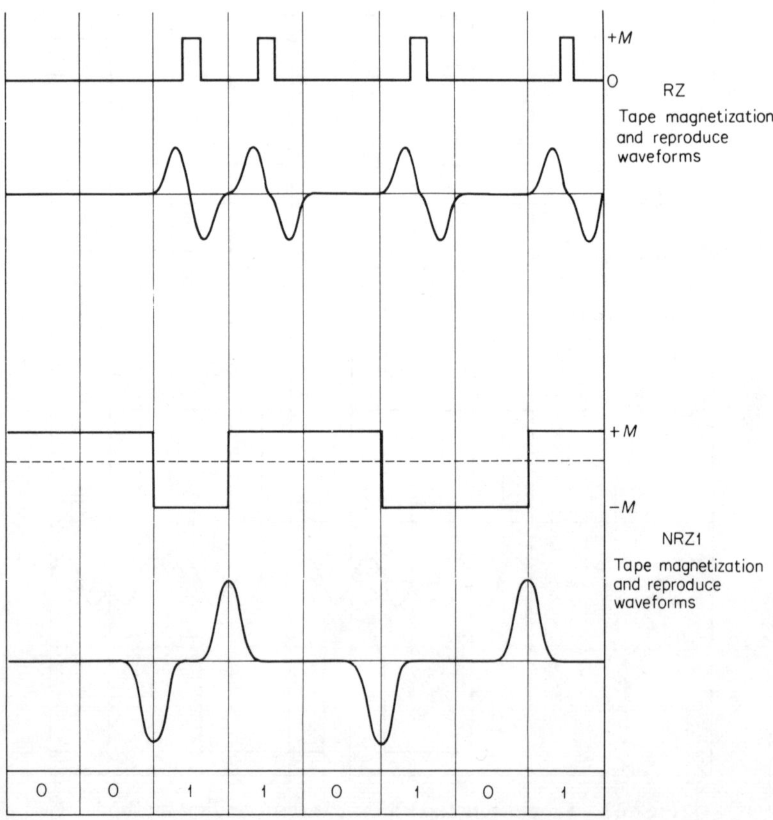

Fig. 5.2. Tape magnetization and reproduce waveforms for RZ and NRZ1.

2. Return-to-bias (RB). The tape is normally saturated. A binary one switches the magnetization in the opposite direction for a small portion of the bit period. The main features of RB are the same as those for RZ, but the reproduce voltage is approximately double because the magnetization excursion extends from saturation to saturation. RB is self-erasing because the bias permits re-recording over old data without a separate erase cycle.

3. Non-return to zero. This system has two basic variants; one is known as NRZ-Change (NRZC) or just NRZ, the other as NRZ-One (NRZ1) or NRZ-Mark (NRZM). In NRZC coding both the recording current and tape magnetization is constant when a series of identical digits – either zeros or ones – follow each other. A transition to saturation in the opposite direction occurs whenever a zero is followed by a one or a one by a zero. The system is not self-clocking and since there is no limitation on the length of an all-zero or all-one sequence, the record/reproduce channel bandwidth must extend to low frequencies. Clearly, NRZ makes better utilization of the medium than RZ as there is a maximum of one transition per bit. The NRZC as described above, has the fundamental disadvantage that if one single bit is in error it causes all successive symbols to change to the opposite until synchronism is regained.

In the NRZ1 coding which is the most widely used coding method in medium density data storage systems, the recording current and hence the tape magnetization is reversed each time a logic one is recorded; no change takes place for a logic zero. Again, there is a maximum of one transition per bit but now if one bit is misread, only that single bit will be in error and following bits are not effected. The system is not self-clocking and since the length of an all-zero sequence is not limited, the channel bandwidth has to be extended to low frequencies. There is no difference between 'no signal' and zero; the coding is sensitive to noise since any noise pulse can be misread for a logic one. There is no restriction on the sequence of zeros and ones, hence when the recording density is increased to the level where there is strong interaction between adjacent pulses there will be a substantial variation in location and amplitude of readback pulses depending on the recorded bit pattern.

In addition to the two basic formats, the NRZC and NRZ1, the non-return to zero coding has many other variants. Some of the disadvantages of NRZ1 can be removed by forcing a flux reversal at every *n*th pulse. This forced flux transition which will be ignored by the data reading channel can be used to re-synchronize the clock at the expense of reduced data density and more complex circuitry.

Investigating now the standard version of NRZ1 with respect to the check list we find that:

(a) Efficiency is high (100%). There is only one flux reversal per bit.
(b) Correlation is poor. The difference between logic zero and one is not the maximum which is possible.
(c) The frequency response needs to be extended to near-zero.
(d) The system is not self-clocking.

(e) Read resolution is good, equal to one-bit period.

(f) The read/write circuits are simple.

(g) Noise sensitivity is fair. There is no difference between logic zero and 'no signal'. Noise pulses may be misread as logic ones.

(h) NRZ1 is poor in respect of peak shift variation, and reproduce amplitude fluctuation at high densities. Error detection capability provided by the code is fair.

4. Phase Encoding. Also referred to as Manchester, Williams, Ferranti method or Phase modulation is currently used in high-density (1000-2000 bit per inch) magnetic tape stores. A cell is occupied – in the idealized schematics of Fig. 5.1 – by a square wave with one cell period. The usual convention is that it represents a logical zero if the first half of the square wave is positive and the second half is negative; for a logical one it is the opposite. Worded slightly differently, a positive-to-negative transition, in the middle of a cell, represents logic zero, and a flux transition from negative to positive, in the middle of the cell, signifies a logic one. If two identical symbols – two logic zeros or two logic ones – follow each other, there is an interbit flux transition at the boundary of the cell. This interbit flux transition will be ignored by the readback circuitry. PE is self-clocking as there is always a flux transition per cell.

Comparing NRZ1 with PE it would appear at first that as phase modulation requires more transitions per bit than NRZ1, it would permit lower maximum bit density. However, there are more important considerations. In NZR1 coding, the spacing of subsequent transitions may vary from a minimum determined by the chosen bit density for a string of logic ones to a large value for a long string of zeros. This random variation of transitions causes peak shift and amplitude fluctuation (see Chapter 4.13). In a phase encoded system, the ratio of maximum to minimum spacing is 2. Hence, although pulse crowding and peak shift phenomena in a system which has fixed magnetic and mechanical parameters is reached at lower bit packing density with PE than with NRZ1 coding, the peak shift and amplitude fluctuation in PE is well defined and constant.

The comparison of the frequency spectra of NRZ1 and PE systems reveals another important difference between the two [3].

The noise spectrum of a practical system (where the term 'noise' includes all unwanted signals, not only random noise) exhibits large peaks at the low frequency end. Although the highest frequency in a PE system is twice that of the corresponding NRZ1, the required band does not extend to low frequencies and a better signal-to-noise ratio can be achieved.

In respect of the coding system's sensitivity to read errors, Fig. 5.1 permits the arrival at some broad qualitative conclusions. If in a NRZ1 coding system a single one or a single zero is misread, it will have no effect on the interpretation of the forthcoming data bits of the sequence. (As NRZ1 is not self-clocking, in a parallel-recording multi-track system each character must have at least one flow-transition, i.e. logic one in it from which the read clock is derived.)

In a typical phase-encoded readback channel the read clock is derived from the data flux transition and the read logic separates the bit flux transitions which occur in the middle of the cell from the interbit transitions at the cell boundaries by opening a 'window' at the expected bit transition time. The possible consequences of a single-bit dropout are illustrated in Fig. 5.3. It is assumed that

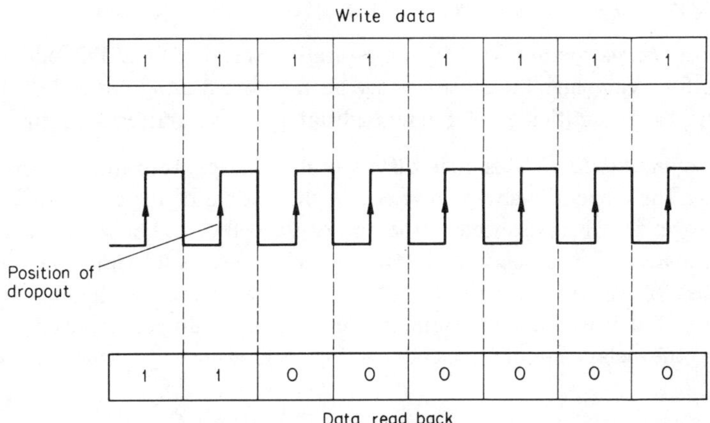

Fig. 5.3. Potential error caused by single-bit dropout in PE.

the dropout occurs in the position of a bit-transition in a long chain of logical one's. The interbit transition following the dropout will now generate the next clock pulse and the subsequent logic one's are misread until synchronism is regained. Thus a single bit-transition dropout can cause an 'error burst' extending over many bit lengths. In actual systems the possibility of this type of error is avoided by provision of a phase-lock oscillator which is normally re-synchronized at every bit flux transition, but can maintain synchronism over a certain number of bits without re-synchronization.

The check list of PE reads as follows:

,(a) Efficiency is low (50%). There are two flux reversals per bit.
(b) Correlation is good; there is the maximum possible difference between logic zero and one.
(c) Bandwidth requirement is fair, does not extend to low frequencies but maximum frequency is twice than NRZ1 at same density.
(d) The system is self-clocking.
(e) Read resolution is poor, limited to half-bit period.
(f) Circuit complexity is higher than for NRZ1.
(g) Noise immunity is good; peak shift and amplitude fluctuation is predictable and constant at any one recording density.

5. Frequency doubling. The basic PE system has many variants. In the Frequency Doubling (also referred to as Pulse Width NRZ1) the logic zeros are represented in the same way as in PE, but the logic ones by a positive or negative saturation level throughout the cell time. For a string of ones there is no flux transition in the middle of the cell. All essential features of this code are the same as those of Phase Modulation. A further alternative (not shown in Fig. 5.1) represents logic ones in the same way as above, but two (or more) square-wave cycles are used to represent a logic zero. Ideally the reproduce waveform consists of large peaks for ones and higher-frequency ripples for logic zeros.

6. Phase change modulation. A non-saturating version of the PE code is shown in Fig. 5.1. Very high bit densities have been reported using this code [7] but at the expense of sacrificing the inherent advantages of saturation recording.

7. Delay modulation. This code differs in its concept from others reviewed up to now. The current is always reversed in the middle of the cell for a logic one. For a logic zero it is reversed at the end of the cell if, and only if, it is followed by another zero. The delay modulation is only one possible implementation of the class of adaptive codes [14, 29]. Adaptive codes are theoretically very attractive but it is doubtful whether the increased circuit complexity is justified. Testing the delay modulation code of Fig. 5.1 against our check list we find that:

(a) Efficiency is 100%; only one flux reversal is required per bit.
(b) Clock pulses are not generated in every cell but maximum distance between flux reversals is 2T.
(c) Read resolution is poor, equal to half-bit period.
(d) Bandwidth requirements are favorable; both high and low frequencies are absent.
(e) Correlation is fair; the two possible states are 'pulse' or 'no-pulse'.
(f) Circuit complexity is high.
(g) Noise immunity is fair; noise pulses can be misread as flux transitions. The separation between flux reversals is either T, 1.5T or 2T.

8. Group codes. The codes reviewed up to now all had the common feature that one binary digit corresponded to one code symbol. In view of the 'two-state' philosophy of digital logic and saturation magnetic recording, this was the obvious choice. However, two or more bits of one data character can be thought to be joined in forming an 'extended digit', and a code symbol assigned to it. For a two-bit extended digit four different waveforms are required and for a 3-bit extended digit the number of different waveforms will be eight.

Let us consider a data byte as a binary coded octal number and construct a code, which conforms to the following rules:

(a) Flux reversals shall not be spaced closer together than in NRZ1 encoding of the binary data.
(b) Flux reversals can take place at the middle or at the boundaries of a cell.
(c) The number of different waveforms must be 8. The complements of the 8 code-symbol waveforms are not considered as distinct waveforms.

This type code is classified by A. E. Whitehouse [27] as a 'single frequency 1 → m group code' because double (or multiple) flux reversals within a cell are not allowed, the spacing between reversals varies between 1 and m bit periods and binary digits are encoded in groups.

Figure 5.4 shows the 8 different waveforms and their complements which are constructed to conform to the given rules.

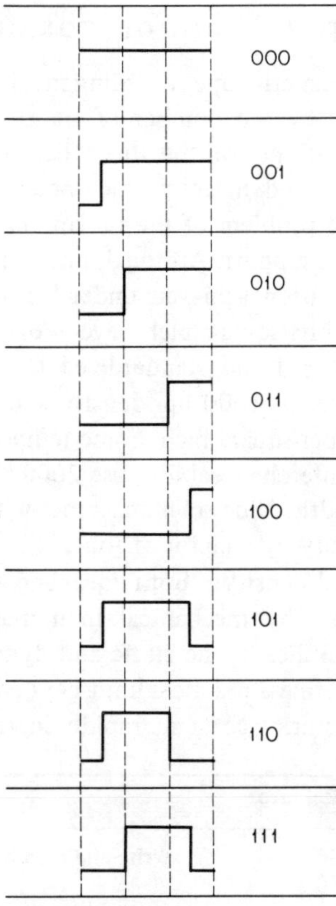

Fig. 5.4. Octal-coded binary group code.

If a sequence of logic zeros is called for, DC magnetization is avoided by selecting the complement of the symbol corresponding to 000 alternately. A sequence of, e.g., octal one followed by octal zero could space the flux reversals $4\frac{1}{2}$ cell lengths, apart; however by optimum selection of a symbol, or its complement, the maximum distance between reversals need only be 3 cell periods. Long runs without flux reversals should be avoided because of clock-synchronization problems. It may be noted that delay modulation can be considered as a group code for which the bit period is equal to one. The number of waveforms is two and both the symbol and its complement are used.

There is an infinite number of ways in which binary digits can be grouped for encoding with various constraints on the reversal intervals. Some group codes offer higher efficiency than NRZ1 however the need for generating and decoding a number of waveforms which are only marginally different from each other and lack of self-clocking does not make them attractive. A systematic study of group codes is beyond the scope of this book.

5.3. PRACTICAL LIMITS OF STORAGE DENSITY

The storage capacity of magnetic tape recording medium can be expressed as bits per lineal inch multiplied by the number of tracks per inch. In Chapter 4 we derived various approximations for the theoretical maximum bit density as a function of the thickness and magnetic parameters of the medium but little attention was paid to the problem of the maximum number of tracks which a specified width of tape can support. Although lineal storage densities as high as 2 x 10^4 bits per inch have been achieved under laboratory conditions and bulk memories with 7 x 10^5 bits/square inch have actually been implemented [1], the current most widely used and standardized formats for information interchange employ a maximum of 1600 bpi density and 9 tracks on half-inch wide tape i.e. 28.8 x 10^3 bits per square inch. Some temporary magnetic tape buffers which do not aim at full interchangeability use 2000 bits/in density but still only 9 tracks per $\frac{1}{2}$ in tape width. Nine-track machines with 6250 bits/in density will increase the storage capacity by a factor of four.

The maximum practical density – both lineal and area density – is in the first instance, determined more by mechanical than magnetic factors. In parallel recording, with multi-track heads, the static and dynamic skew, head, tape and transport imperfections define a practical limit beyond which reliability becomes questionable and manufacturing costs are rapidly increasing. Figure 5.5 tabulates

Source	Typical error	Remark
Gap scatter	2 x 80	Write and read head
Tape skew	2 x 50	
Write time error	50	Caused by variation in impedance between two halves of one head
Peak shift	100	Worst-case pattern
Miscellaneous	30	
Total	440	

Fig. 5.5. Representative worst case bit displacement on a 800 bit/in NRZ1 channel (nominal bit spacing 1250 μ in). All dimensions in μ in.

representative figures for the displacement of the position of the pulse* in an 800 bits/inch density medium-speed NRZ1 coded system. Without skew compensation the worst-case total error can amount to about $\frac{1}{3}$ of the nominal

* From its ideal position.

distance between pulses in a typical medium-speed digital recorder. Some of the mechanical deficiencies such as static skew caused by gap scatter can be relatively easily corrected by electronic means, others such as reducing tape motion irregularity, below a certain level, become increasingly difficult and expensive. We have reviewed such limiting factors as head-to-tape separation at high speed, peak shift and pulse crowding, and write time asymmetry (variation between the impedances or two halves of a centre-tapped head and from track-to-track). Other fundamental limiting factors are noise, cross-talk in multi-track heads and the same between recorded tracks.

The term 'noise' has a somewhat different interpretation in analogue and digital systems. In audio and instrumentation recording the tape and amplifier noise represents one of the fundamental performance limiting factors, and consists of such components as noise from the fully erased tape, modulation noise which is proportional to the signal level, tape noise caused by high-frequency bias (which is higher than that of bulk-erased tape) [16, 17, 18, 19, 20, 21, 26] and finally, amplifier and thermal noise. In digital recording where the normal reproduce level from a saturation-to-saturation transition even at densities where pulse crowding is substantial, is still much higher than the tape noise and the 'average' noise from all other sources. More significant are the random pulses caused by surface irregularities, drop-ins and the interaction between heads and tracks. The maximum track density, the track width and inter-track separation is determined – in addition to mechanical factors – by cross-talk and interaction between heads and tracks. The reproduce head output, all other factors being constant, is nearly linearly related to the track width, hence there is a minimum width below which the signal-to-noise ratio becomes unacceptably low. Also, increase in track width can compensate for loss of signal amplitude at high lineal densities. The track separation is – ignoring the mechanical accuracy problems – primarily dictated by signal-to-noise deterioration with decreasing track separation caused by pickup from adjacent tracks and cross-talk in multi-track heads. The adjacent track interaction can marginally shift the position of the recorded transition [24]. The density limiting factors are different on discs and tapes; moving head discs which have a single head traversing the recorded tracks tolerate much higher track density than tapes so that proposed disc standards specify area densities in excess of $100 \cdot 10^3$ bits/in^2.

Reliability analysis of magnetic tape storage systems leads to the conclusion that, ignoring mechanical elements, the most likely source of data errors is the magnetic head/tape interaction area. Typically, the source of error (ignoring catastrophic failures and assuming the use of 'certified' computer tape which is free from permanent errors) is a minute particle of dust, loose oxide, which lifts the tape away from the head. Errors may also be caused by statistical fluctuations. It has been shown [10] that to achieve an error probability better than 1 in 10^{11}, the signal/noise ratio must be better than 20 dB. Thus for a recording material of given coercivity, remanence and thickness the minimum track width is determined by the need of achieving a signal/noise ratio in excess of 20 dB at the lowest tape speed for which the machine is intended. Digital recorders which employ electronic error correction and detection can achieve an error rate better than 1 in 10^{10} bits at 1600 bits/in density if operated under optimum conditions (air-conditioned dust-free clean room, carefully adjusted and maintained transport, and a computer grade tape, free of permanent errors) and claims for even 1 in 10^{12} have been made.

5.4. STANDARDIZATION OF RECORDING FORMATS

An important step towards improving interchangeability of magnetic tape between various installations has been achieved in the late sixties by standardization of the key recording parameters. The format used in IBM (International Business Machines) has become the *de facto* recording standard before various national and international bodies (American National Standards Institute, European Computer Manufacturers Association, International Standards Organization) have issued their recommendations. The ANSI and ECMA standards are based on the IBM format and are identical in all but some minor points such as block length, interblock gap tolerance, terminology, etc. Our purpose here is to give a brief summary of the salient features of the formats which are currently widely used in the computer industry. It is important to recognize that computer installations which serve different purposes, have different architecture and technology; programming techniques use magnetic tape data stores which conform to the same recording format standards. This situation differs from the mainframe memories where standardization has not reached the same stage, but of course the problem of interchangeability does not arise.

As recording materials, heads, transports and electronics improve and new recording techniques are introduced the standards are revised and updated, thus the standards reflect the state of the art but with some delay. At the same time, as large investments have been made in installations which conform to current standards, the change to more advanced ones, if they are not compatible with the older formats, is naturally rather slow.

Recording of parity

Data which are transmitted between the central processor and the magnetic tape system must be checked for errors; in particular it must be established whether the data sent from the central processor were correctly recorded on the tape and whether they were correctly recovered.

One of the simplest yet most widely used error checking systems is based on appending a single redundant bit to every data character. The redundant bit indicates whether the total number of information bits in the message is even or odd. Either convention may be used, but if odd parity is chosen (i.e. the total number of logic ones in the character including the redundant bit, which itself is a logic one, is made to be odd) the message will contain at least a single one. Thus a character of all zeros (which represents an even number) is distinguishable from 'no message'. In an NRZ1 coding system without an external clock there must be at least one flux transition (logic one) in every character, otherwise the character cannot be recovered. Hence in NRZ1 coding systems the number zero is represented by a combination of digits which contains a logic one, or if an all-zero character is permitted, odd parity must be used.

Should an error occur, the sum of the digits will no longer be odd, (or even) thus the error will be indicated. Only such errors will be detected which violate the system's parity convention. During recording the controller logic generates and records the appropriate parity bit for each recorded character on one track of the tape which is reserved for this purpose.

The parity bit may be used for error detection and correction in a variety of ways. The readback logic may simply check each character parity whether it conforms to the system convention or regenerate the parity bit from the data bits alone and compare this with the parity bit read off the tape. If an error is found the central processor may request re-reading the block of data several times and if the error has not been cleared, reject the block. A more detailed explanation of error detection and correction techniques follows in Chapter 6.

Seven-track tape format

The choice of seven tracks on a half-inch wide tape was dictated by the organization of the earlier computers which used six bit characters as the basic unit of data transmission. Six-bit codes (such as the BCD and six-bit ASCII code) permit $2^6 = 64$ different bit combinations to represent the numbers and letters of the alphabet. The seventh track is reserved for the character parity bit (often referred to as lateral parity, frame parity or vertical parity check, VRC). The track width of the write head is chosen slightly wider than that of the read head. The recording method is NRZ1 with 200, 556 and 800 bits/in density.

At the end of the block a further check character, the Longitudinal Redundancy Check character is written. This consists of the collection of parity bits which are generated separately for each data track in the block. The frame parity bit can be chosen to be either odd or even but the LRC parity bit is always even, i.e. it makes up the total number of ones in one track of the data block to be even. The LRC character is written four bit spaces from the end of the data block or tape mark character and followed by the 0.75 in long interblock gap. The tape mark consists of a single character which has logic ones in track 1, 2, 3 and 4 and zeros in all others, plus the LRC character. Assigning the binary weight 1, 2, 4, 8 to the tracks 1 to 4, the numerical value of the tape mark will be 15 decimal or 17 octal. The LRC character for the tape mark will be identical with the tape mark itself.

Figure 5.6 also shows the load point and end-of-tape reflective markers which indicate the beginning and end of the usable portion of the tape. The markers are on the non-coated side of the tape.

The tape is saturated in the erased direction in the inter-block gap area. The erase polarity is defined in such a way that the north magnetic pole of any erased section shall point in the direction of the beginning of tape (load point marker).

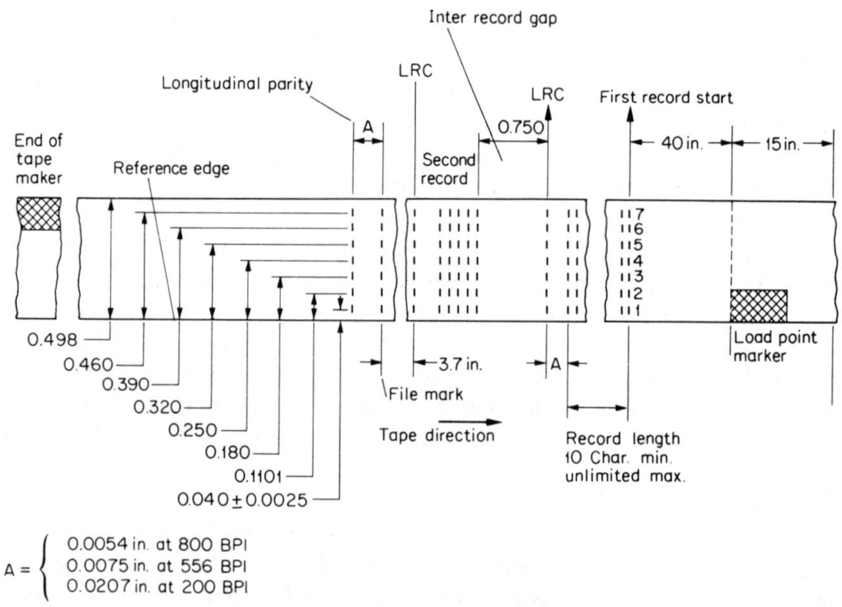

Fig. 5.6. Seven track tape format.

Nine-track tape format

The use of nine tracks permits a convenient data organization for the computers which are byte-orientated i.e. the basic unit in which data are manipulated and transmitted is an 8-bit byte. As in the seven track format, one track is reserved for character parity. The method of recording is NRZ1 for 200, 556 and 800 bits/in density and phase-encoded for 1600 bits/in. The dimensions and format are shown in Fig. 5.7; the various standards agree on all critical dimensions but are marginally different in some details.

In nine-track NRZ1. (Fig. 5.7) format, the character parity is always odd. The parity track is track No. 4 (BS. 5403 track numbering) which improves the parity track reliability; the likelihood of errors is higher in the tracks near the edge of the tape than in the centre. The data block is followed by a cyclic redundancy check (CRC) character (see Chapter 6) and terminated with the longitudinal redundancy check (LRC) character. The LRC permits checking of parity in each track in the block in longitudinal direction and restores all tracks to the DC erase polarity.* The tape is saturated in the 0.5 in long interblock gap

* The LRC character has always odd parity.

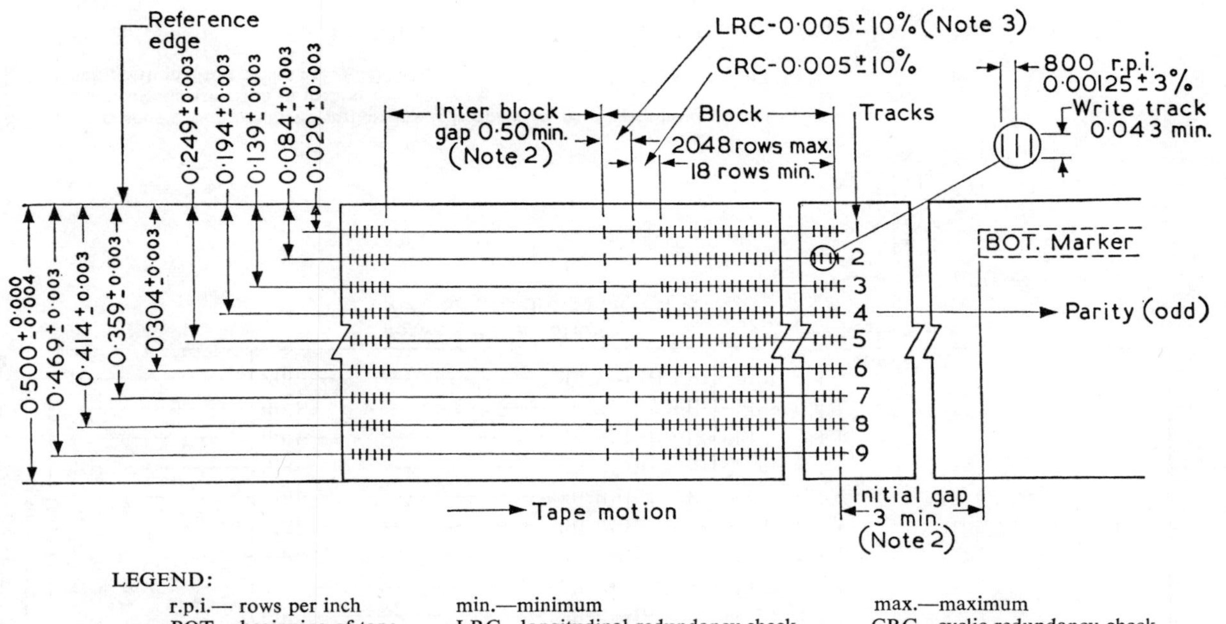

LEGEND:

r.p.i.— rows per inch	min.—minimum	max.—maximum
BOT.—beginning of tape	LRC—longitudinal redundancy check	CRC—cyclic redundancy check

NOTE 1. Tape is shown with oxide side up, Read/Write head on same side as oxide. Tape shown representing locations of 1 bits in all tracks, NRZ1 recording; 1 bits produced by reversal of flux polarity, tape fully saturated in each direction.

NOTE 2. Tape to be fully saturated in the erased direction in the inter-block gap and the initial gap.

NOTE 3. A longitudinal check bit is written in any track if the longitudinal count in that track is odd. Parity is ignored in the longitudinal redundancy check row.

NOTE 4. Parity of cyclic redundancy check row is odd if an even number of data rows are written; it is even if an odd number of rows are written.

NOTE 5. All dimensions are given in inches.

Fig. 5.7. Nine track tape format, NRZ1 (after B.S. 4503).

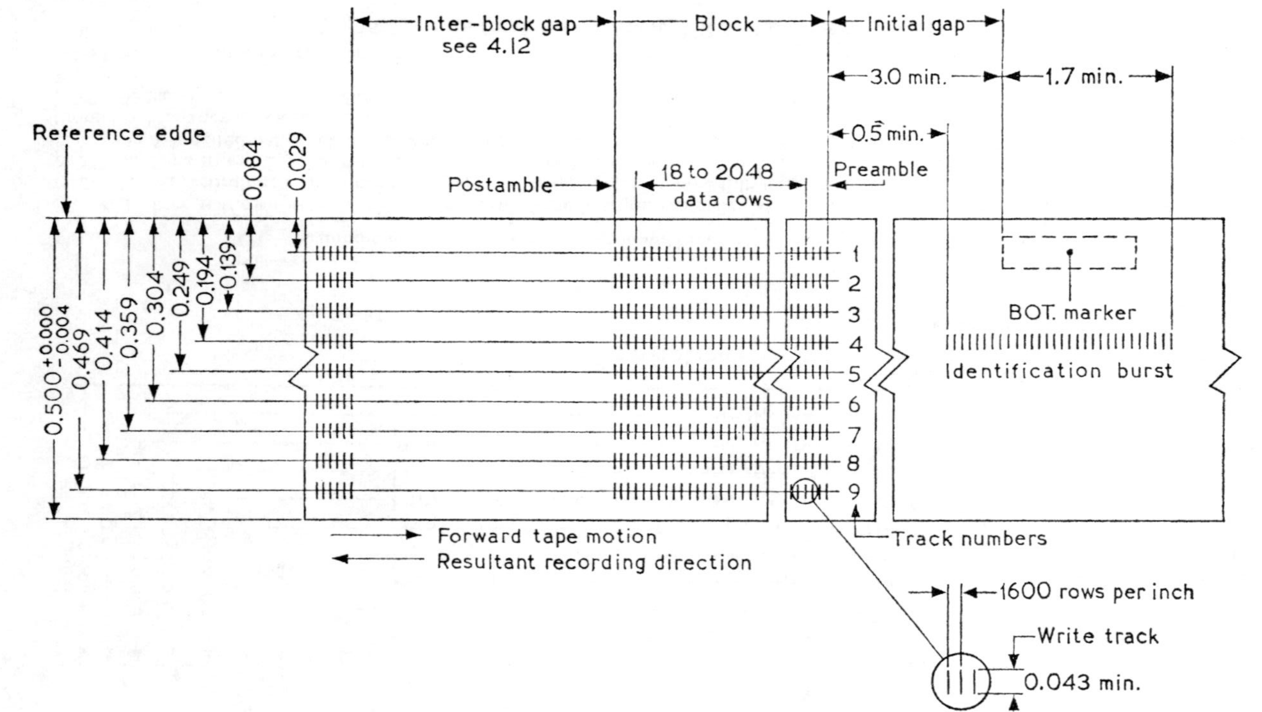

NOTE 1. Tape is shown with magnetic surface towards observer, read-write head on same side.
NOTE 2. All dimensions are given in inches.
NOTE 3. Track positional tolerances are ±0.003 in.

Fig. 5.8. Nine track tape format, PE (after B.S. 4503).

in such a direction that its north magnetic pole points in the direction of the beginning of the tape (load point marker). The tape mark character consists of a one in tracks 2, 3 and 8.

The nine-track phase-encoded. (Fig. 5.8) format does not utilize the CRC and LRC characters for error checking. Each data block is preceded by 40 all-zero bytes followed by a single all-ones byte (preamble) and is terminated by a single all-one byte followed by 40 all-zeros bytes (postamble). The preamble synchronizes the read detection circuits so that ones and zeros are correctly identified in the data bytes which follow. The symmetry of preamble and postamble permits reading in either direction. At the load point (beginning of tape) an 'identification burst' consisting of 1600 flux reversals per inch, is written in the parity track. Odd character parity is specified. The Tape Mark consists of 64 to 256 flux reversals at 3200 flux changes per inch in tracks 1, 2, P, 5, 7, 8. The remaining tracks are DC erased.

5.5. DATA RECOVERY

The readback electronics has to establish whether the signal reproduced during a cell period (or distance occupied by one cell along the tape) represents a logic zero or a logic one. The time span available for decision was recognized as an important feature of the coding system. Consider as an example the NRZ1 coding illustrated in Fig. 5.1. The write current flows at all times to cause positive or negative saturation of the medium and changes whenever a logic one is recorded. The reproduce voltage, which is a time derivative of the flux change, appears as a discrete pulse at the flux transition times with its polarity depending on the direction of the transition for a logic one and as 'no signal' for a logic zero. An amplitude-detection circuit works on the principle that a signal is classified as logic one if its amplitude exceeds a threshold level at the sampling time, and logic zero if it is below it.

The simple amplitude-sensing circuit, as illustrated in Fig. 5.9, has serious limitations. The reproduce voltage amplitude, even in saturation type recording, exhibits considerable fluctuations because the sensitivity of the reproduce signal level to the head-to-tape separation and variation of the surface properties of the recording media. Further, at high densities, the reproduce amplitude is a function of the bit sequence as illustrated in Fig. 4.13. As amplitude sensing introduces an amplitude-dependent shift in the detection point, it is limited to low-density non-critical applications.

The disadvantages of the amplitude-detection are eliminated by the peak detection which is the most frequently used detection method in high-density systems. The peak of the reproduce pulse waveform is detected by differenti-ation, followed by an amplifier-limiter, a low-pass filter, or integrator. The principle is illustrated in Fig. 5.10. The zero-crossings which give the location of

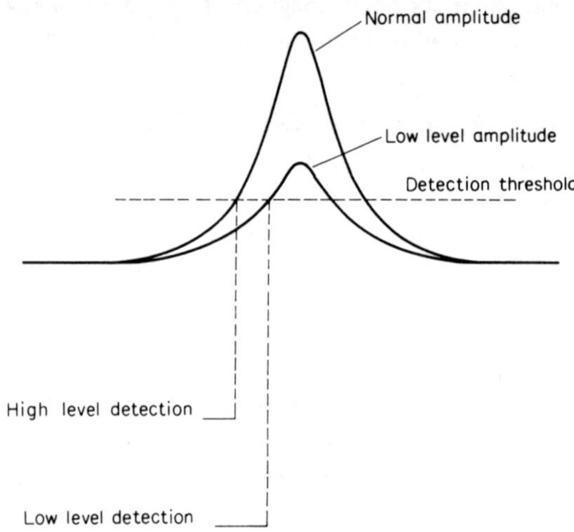

Fig. 5.9. Amplitude detection.

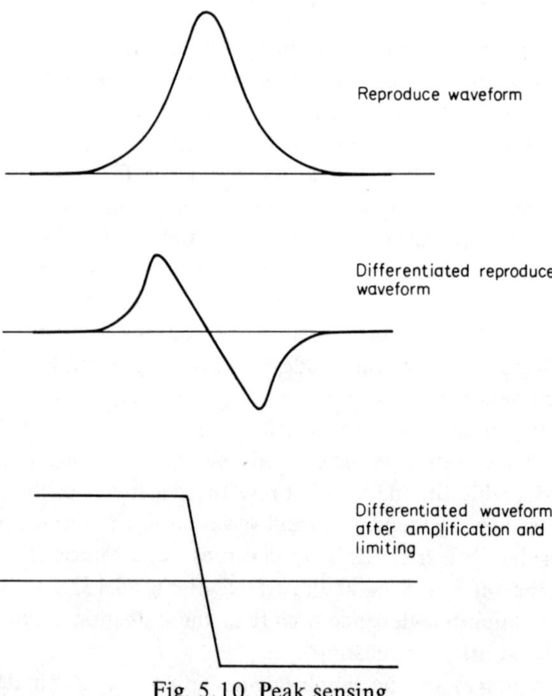

Fig. 5.10. Peak sensing.

the peak are, to a large extent, independent of the amplitude of the reproduce
waveform and can be sensed by standard pulse circuit techniques; a solution
using second differentiation is illustrated in Fig. 5.11.

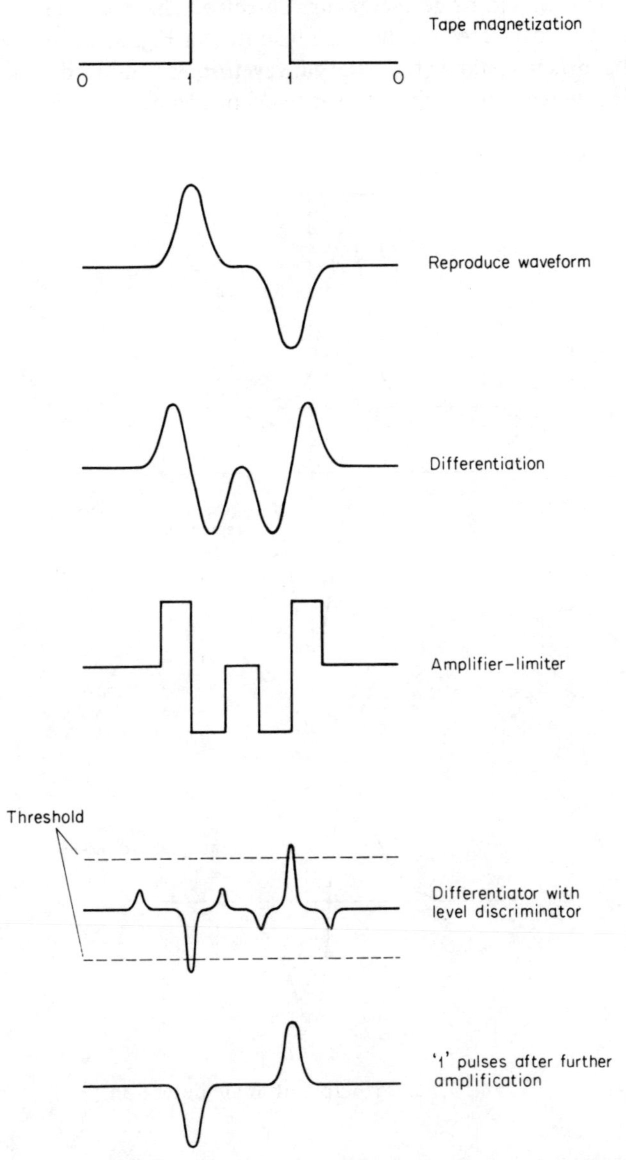

Fig. 5.11. Schematic of data recovery with double differentiation.

Many alternative data recovery systems, which are known in the literature as 'differencing', 'cancelling', or 'angle detection', are based on delaying the reproduce signal by a specified amount and then adding or subtracting the delayed signal from the original. The difference signal is amplified, and amplitude-limited, and its zero-crossings detected. The principle of angle detection, which is widely used on discs is illustrated in Fig. 5.12. In a variant of the delay and subtract method, the delayed waveform is combined with the original in an AND gate when it is referred to as the coincidence method.

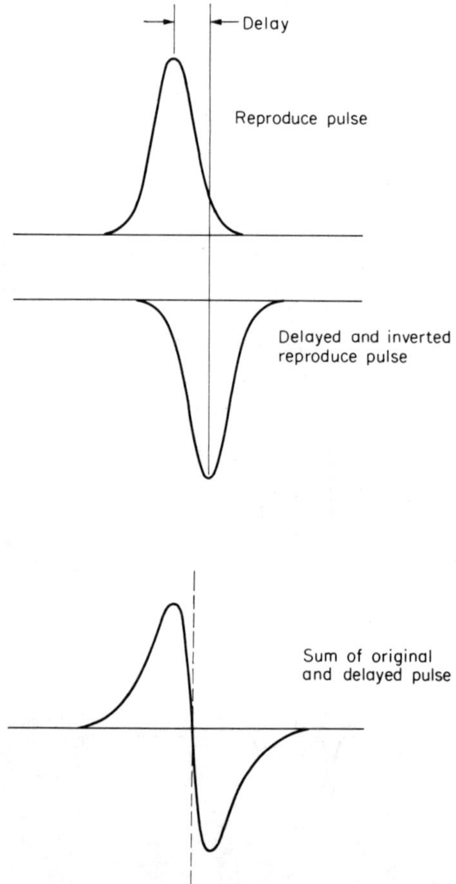

Fig. 5.12. Principle of angle detection.

In phase-encoded systems there is, contrary to NRZ1 coding, a continuous readback signal which, because of the bandwidth limitation of the record/reproduce chain, and the inherent properties of the recording media appear as a near-sinusoidal signal. In high-density phase encoded systems, the zero-crossings of this quasi-sinusoidal signal can be used for accurate location of the readback pulse. This detection system is used on some low-cost system in particular on single-track serial recording.

5.6. READ/WRITE CIRCUITS

The readback channels dealing with low-level signals are more complex than the write circuits. Figure 5.13 shows in the form of a block diagram the essential elements of a readback channel for recovery of the phase encoded data. As always, in digital logic and circuit design, the engineering solutions will take a large variety of forms so that it is hardly possible to define a 'typical' read channel. Detailed circuit analysis and design is beyond the scope of this book.

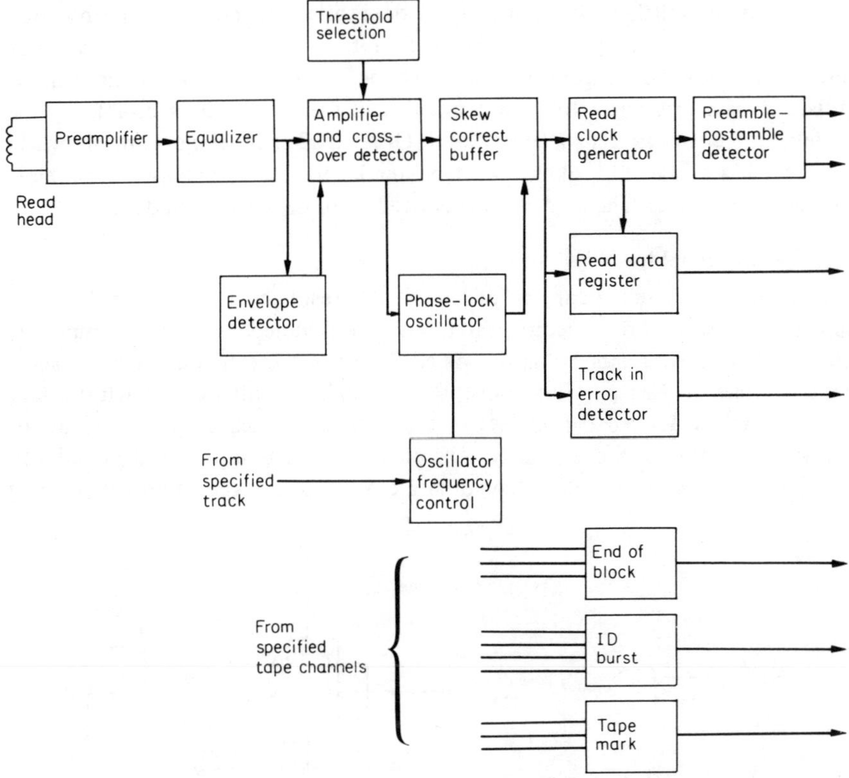

Fig. 5.13. PE readback channel functional elements.

The read head feeds the low-level readback signal into the preamplifier, often mounted in the vicinity of the read head. The preamplifier may be followed by an amplitude and phase corrector (equalizer). It is generally necessary to suppress noise by introducing a threshold stage which permits the passage of signals, only if they are above the threshold level. The threshold is set higher in read-after-write mode than in read only mode. Sharp noise pulses of short duration may exceed the threshold level, hence the peak detector may be followed by an integrator and the amplifier bandwidth is kept at the minimum, consistent with the requirements of adequate pulse resolution. The readback channel may contain a rectifier so that both positive-going and negative-going signal excursions are presented to the amplifier-detector as similar polarity signals. Each channel has an independent skew correction buffer which may have as many as four stages in serial and an independent phase-locked oscillator. The oscillator is synchronized at the beginning of the block by the preamble, and may receive from one or more channels on the tape, a continuous signal derived from the rectification of the pulses on that channel representing the average tape speed. The phase-locked oscillator opens a 'window' at the expected data bit time thus permitting the transfer of data bits but not the interbit flux transitions. If a data bit is lost because of tape dropout and the phase-lock oscillator is not re-synchronized, it still may be able to maintain synchronism for adjacent data bits but errors are counted, and the track will be disabled if the number of errors exceeds a preset level. Other read-channel elements shown in Fig. 5.13 are the end of block, tape mark, ID burst, preamble/postamble recognition circuits. The read clock is derived from the de-skewed data.

Readback channel for NRZ1 system

The main functional elements of an NRZ1 readback channel are the pre-amplifier, the threshold selector, the rectifier-peak detector, the main amplifier, the de-skew register and the buffer. As NRZ1 is not self-clocking, each character must contain at least one flux transition, i.e. logic one bit from which the data clock signal is derived for all bits in the character. Such systems exclude an all-zero character with even parity. The outputs from all tape tracks are fed into an OR gate from which the delayed data clock signal is generated for strobing

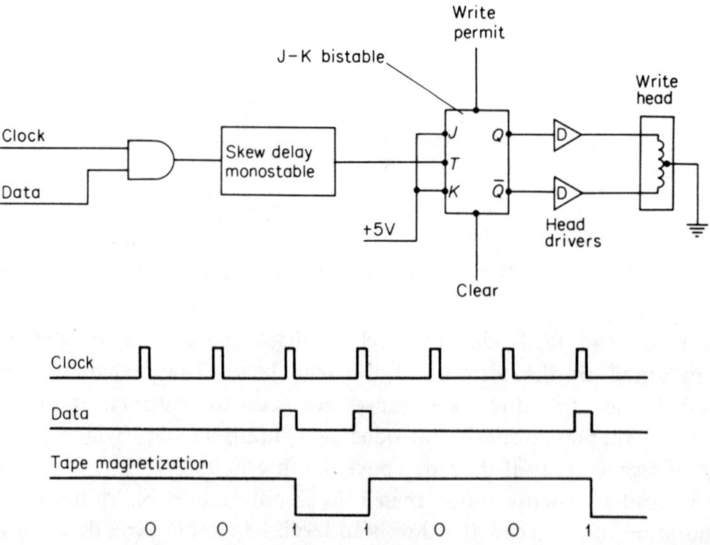

Fig. 5.14. Simplified NRZ1 write circuit.

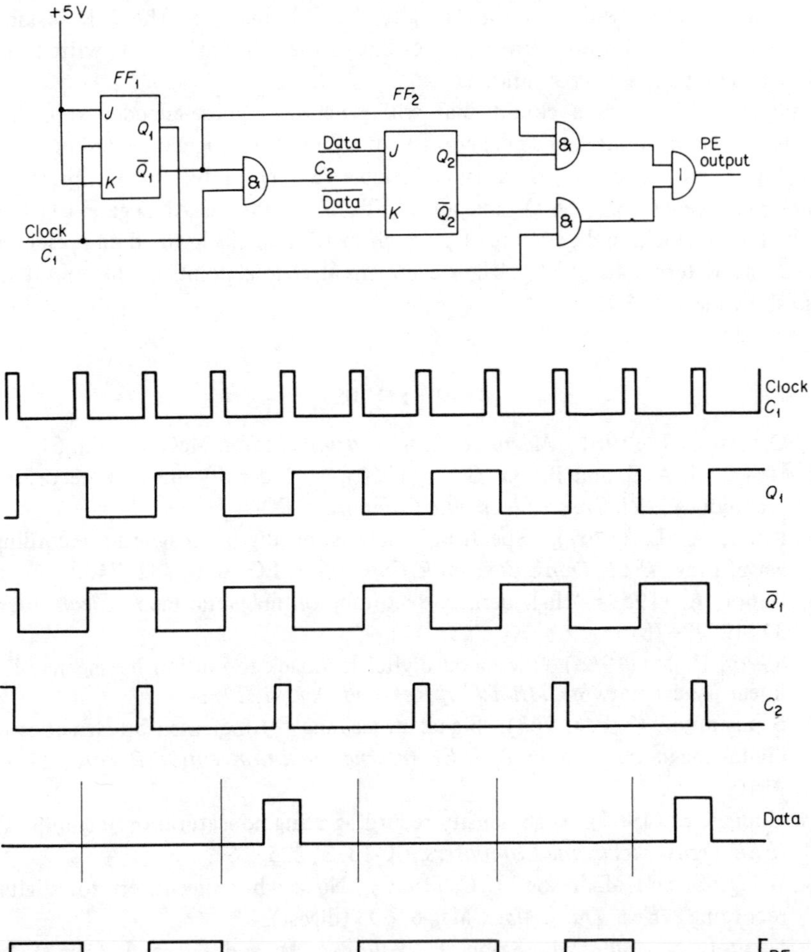

Fig. 5.15. Simplified phase encoder.

the skew-corrected data into the output buffer register. The reproduce channel may contain the read parity generator, read/write parity comparator, CRC and error pattern registers with associated circuitry as described in Chapter 6. Data reliability is enhanced by employing a 'low' and a 'high' threshold level simultaneously, and data is processed only if it passes both thresholds.

Write circuits

A simplified block diagram of one possible solution for an NRZ1 write circuit is shown in Fig. 5.14. Data gated with a clock, trigger the adjustable write skew delay monostable, which in turn toggles the J–K bistable. The J–K bistable energizes, via the D power drivers, either one or the other half of the write head, so that the tape is always saturated.

Figure 5.15 shows a circuit that will generate a phase-encoded recording current when the input data signal consists of a pulse for a logic one, and no pulse for logic zero. The FF_1 J–K type bistable is toggled by the C_1 precision clock-pulse generator. The $\bar{Q}_1$ output of FF_1 is fed into an AND gate, together with the C_1 clock, and generates C_2 which in turn clocks in the data present at the J and K terminals of FF_2. The waveforms at various points of the circuit are also shown in Fig. 5.15.

REFERENCES

1. Davies, G. L. (1961), *Magnetic Tape Instrumentation,* McGraw-Hill, 51–72.
2. Hoagland, A. S. and Bacon, G. C. (1960), 'High density magnetic recording techniques', *IRE Trans. Electronic Computers,* **EC-9**, 2–11.
3. Knoll, A. L. (1967), 'Spectrum analysis of digital magnetic recording waveforms', *IEEE Trans. Electronic Computers,* **EC-16**, 6, 732–743.
4. Gabor, A. (1959), 'High density recording on magnetic tape', *Electronics,* **32**, 42, 72–75.
5. Sierra, H. M. (1963), 'Increased digital readback resolution by means of a linear passive network', *IBM J. of Res. and Dev.,* 7, 22–33.
6. Schlaepfner, C. E. (1963), 'Signal processing for increased bit densities in digital magnetic recording', *IEEE International Convention Record,* **11**, **4**, 2–10.
7. Hopner, E. (1964), 'High density recording using nonsaturation techniques', *IEEE Trans. Electronic Computers,* **EC-13**, 3, 225–261.
8. Wong, E. and Mallinson, J. C. (1970), 'Noise-whitening filters for digital recording', *IEEE Trans. Mag.,* **Mag-6**, 603 (digest).
9. Damron, S., Miller, J., Salbu, E., Wildman, M. and Lucas, J. (1968), 'A magnetic random access terabit memory', Fall Joint Computer Conference, AFIPS Conf. Proc., **33**, II, 1381.
10. Mallinson, J. C. and Ragosine, V. E. (1971), 'Bulk storage technology: magnetic recording', *IEEE Trans. Mag.,* **Mag-7**, 3.
11. Skov, R. A. (1958), 'Pulse time displacement in high density magnetic tape', *IBM J. of Res. and Dev.,* **2**, 130.
12. Eldridge, D. F. and Baaba, A. (1961), 'The effects of track width in magnetic recording', *IRE Trans. Audio,* **9**, 1, 10–15.
13. Ashlock, J. C. (1971), 'Transversal filter for high-density recording', *IEEE Trans. Mag.,* **Mag-7**, 3, 421–422.
14. Kobayashi, H. (1971), 'New coding approach to digital magnetic recording', *IEEE Trans. Mag.,* **Mag-7**, 3, 422–423.
15. Ziman, G. C. (1971), 'Maximum pulse packing density on magnetic tape', *Electronic Engineering,* **34**, 521.

16. Eldridge, D. F. (1964), 'DC and modulation noise in magnetic tape', *IEEE Trans. Audio,* **Au-12,** 5, 100–104.
17. Hung, J. W. (1960), 'Transfer function and error probability of a digital magnetic tape recording system', *J. Appl. Phys.,* **31,** 369–379.
18. Stein, I. (1962), 'Analysis of tape noise', *IRE International Convention Record,* **10,** 7, 42–65.
19. Mee, C. D. (1964), *The physics of magnetic recording,* North-Holland Publishing Co., Amsterdam, 127–134.
20. Eldridge, D. F. (1963), 'A special application of information theory to recording systems', *IEEE Trans. Audio,* **11,** 1, 3–6.
21. Mallinson, J. C. (1969), 'Maximum signal/noise ratio of tape recorders', *IEEE Trans. Mag.,* **Mag-5,** 182–186.
22. Davies, A. W. (1967), 'Effects of writing process and crosstalk on timing accuracy of pulses in NRZ digital recording', *IEEE Trans. Mag.,* **Mag-3,** 217–222.
23. Richards, R. K. (1957), *Digital computer components and circuits,* Van Nostrand, 314–353.
24. Katz, A. S. (1964), 'Space-borne recorder triples packing density', *Electronics,* **37,** 84–88.
25. Hoagland, A. S. (1960), 'A logical reading system for non-return to zero magnetic recording', *IRE Trans. Electronic Computers,* **EC-4,** 1, 2–11.
26. Byers, R. A. (1971), 'Theoretical signal to noise ratios for magnetic tape heads', *IEEE Trans. Mag.,* **Mag-7,** 2, 254.
27. Whitehouse, A. E. (1970), 'Aspects of magnetic recording', Ph.D.Thesis, The University of Manchester.
28. Dunstan, E. M. and Whitehouse, A. E. (1967), 'A 2000 bits-per-inch magnetic drum channel with high noise immunity', *IEEE Conference on Computer Technology, Manchester.*
29. Gabor, A. (1967). 'Adaptive coding for self-clocking recording', *IEEE Trans. Electronic Devices,* **EC-10,** 866–868.
30. Smaller, P. (1965), 'Reproduce system noise in wide-band magnetic recording system', *IEEE Trans. Mag.,* **Mag-1,** 44–47.

6 Error detection and correction

Reliability of data recorded on magnetic tape can be greatly enhanced by the use of error detection and correction electronics.

The most widely used error detection scheme is based on the parity check. It is common practice in computer technology to generate the parity for a character whenever it is transferred from register to register, or from register to a peripheral. In magnetic data recording, a parity bit is generated for each data character and recorded on an additional channel; further, a parity character can be generated for a block of data. The read and write parities are compared and any discrepancy registered as an error. By use of the character parity and block parity, the bit in error can be located and, under some circumstances, corrected.

In this chapter we want to discuss the generation and use of the Cyclic Redundancy Check character for error detection, and correction on nine-track tape. Nine-track tape standards such as B.S. 4503 refer to the CRC character and give the rules of manipulation without any background information. The general theory of digital codes and the investigation of the error-detecting and correcting properties of codes is beyond the scope of this book; we refer to standard textbooks on the subject in particular References [1, 3, 9, and 10]. However, a brief introduction to cyclic codes is necessary. The objective is to illustrate the statements with examples instead of rigorous proofs.

6.2. ERROR CHECKING ON SEVEN- AND NINE-TRACK TAPE – NRZ1 RECORDING

The addition of one redundant bit – the parity bit – to each character permits the detection of single-bit errors. If a further longitudinal parity character, containing the parity bits generated for each track is added, the check system will permit the detection of single, double and triple errors in the block (or, as this follows from the theory of error checking codes, correct single and detect double errors). This can be easily verified by direct tests.

A more complex error detection and correction system which uses the Cyclic Redundancy Character for correction of certain types of single-track errors, is based on the properties of *cyclic codes* which we will now briefly introduce. The data recorded on tape consists of groups of logic zeros and ones. A convenient way of representing such binary characters (or messages in the terminology of communication theory) is to consider each bit as the coefficient of a dummy variable x. This will permit algebraic manipulation of the message. The *degree* of this polynomial of x is equal to the exponent of the highest power of x in a term with a non-zero coefficient. Polynomials are added bit-by-bit according to the rules of binary arithmetic *but without a carry* (modulo-2 addition). Multiplication and division in modulo-2 arithmetic follows the rules of conventional binary arithmetic. As a simple example, consider two binary messages: (Example 1)

$$A = 1\ 1\ 0\ 1\ 0\ 1 = x^5 + x^4 + x^2 + x^0, \tag{6.1}$$

$$B = \quad 1\ 0\ 1\ 1\ 1 = x^4 + x^2 + x^1 + x^0, \tag{6.2}$$

The modulo-2 sum of A and B is

$$A \oplus B = 1\ 0\ 0\ 0\ 1\ 0 = x^5 + x^1. \tag{6.3}$$

As modulo-2 addition is identical with the 'exclusive-or' Boolean operation, the symbol $\oplus$ is often used instead of $+$ sign.

It should be noted that modulo-2 addition and subtraction produces the same result. The somewhat surprising, but correct conclusion is that in modulo-2 arithmetic $-1 = +1$.

Prior to storing the message on the tape it will be encoded. In this chapter the word code, encoding, or decoding will be used in the same sense as in the communication theory and engineering. A useful method of encoding a message is to multiply it (modulo-2) by a suitably chosen generator polynomial. The accuracy of a message will then be checked – at the receiving end – by dividing it by the same generator polynomial; if the message was received correctly, the division will give a zero remainder.

The multiplication of the $M(x)$ message polynomial of $k - 1$ degree with a $G(x)$ generator polynomial of $n - k$ degree gives an encoded message word which will contain n total number of bits from which k will be information bits and $n - k$ will be check bits. A more convenient form of the code word will be obtained by first multiplying the message polynomial by x^{n-k}, the highest degree term of the generator polynomial, then dividing by $G(x)$ and addition of the remainder to $M(x)x^{n-k}$. The division of $M(x)x^{n-k}$ by $G(x)$ will give, in general, a quotient $Q(x)$ and a non-zero remainder, $R(x)$, or

$$M(x)x^{n-k} = G(x)Q(x) + R(x). \tag{6.4}$$

Because we use modulo-2 arithmetic, we can write

$$M(x)x^{n-k} + R(x) = G(x)Q(x). \tag{6.5}$$

The left-hand term in Equation (6.5) is evenly divisible with the generator function, and the degree of the remainder is less than $n - k$. This left-hand side of Equation (6.5) forms a code word in a *systematic code* where the k highest order coefficients in the code word are the same as the message coefficients, and the $n - k$ lower order coefficients are the check bits. In other words, the message appears in its original form in the code word and the check bits are appended to it. The detection of an error is accomplished by dividing the received code word by the generator polynomial; the correct message is evenly divisible without a remainder. The codes derived in the above manner are 'cyclic' because when a word in a cyclic code is shifted by any number of positions in such a manner that the bits which are 'shifted out' on one end are entered on the other, another valid word of the code is formed. As an illustration let us consider the $(n, k) = (7, 4)$ code, with the generator function $G(x) = x^3 + x^2 + 1$. The $(n, k) = (7, 4)$ code has a total number of $n = 7$ bits from which $k = 4$ bits are information, and $n - k = 7 - 4 = 3$ bits are check bits. Suppose that the original message to be encoded is (Example 2)

$$1\ 1\ 1\ 0. \tag{6.6}$$

To encode this message we first multiply the corresponding polynomial with x^{n-k}, divide by $G(x)$ and add the remainder of the division to $M(x)x^{n-k}$. Note that we do not need the quotient of the modulo-2 division, only the remainder; this can be obtained by writing the highest order term of the divisor under the highest term of the dividend and carrying out modulo-2 addition. The process is repeated until the remainder is of lower order than the divisor.

The multiplication gives:

$$(x^3 + x^2 + x^1)x^3 = x^6 + x^5 + x^4. \tag{6.7}$$

The R remainder is obtained by modulo-2 division of Equation (6.7) by $G(x)$:

$$
\begin{array}{l}
1\ 1\ 1\ 0\ 0\ 0\ 0 \\
\underline{1\ 1\ 0\ 1} \\
\qquad 1\ 1\ 0\ 0\ 0 \\
\qquad \underline{1\ 1\ 0\ 1} \\
\qquad 0\ 0\ 0\ 1\ 0 = R.
\end{array}
\tag{6.8}
$$

Thus the coded message will be:

$$F = 1x^6 + 1x^5 + 1x^4 + 0x^3 + 0x^2 + 1x^1 + 0x^0. \tag{6.9}$$

$$\underbrace{\qquad\qquad\qquad}_{n \text{ information bits}} \quad \underbrace{\qquad\qquad}_{n - k \text{ check bits}}$$

If the message arrives distorted, it can, in general, be represented by

$$F'(x) = M(x)x^{n-k} + R(x) + E(x), \tag{6.10}$$

where $E(x)$ represents the error pattern. The correctness of the message is tested by division of $F(x)$ by $G(x)$. As in Equation (6.10) $M(x)x^{n-k} + R(x)$ is evenly divisible by $G(x)$ – we constructed it so that it should be – only $E(x)$ needs to be tested. Suppose now that in the above example where the correct coded message is

$$F = 1\ 1\ 1\ 0\ 0\ 1\ 0, \tag{6.11}$$

the two highest order bits get distorted and the received message is

$$F = 0\ 0\ 1\ 0\ 0\ 1\ 0. \tag{6.12}$$

The error pattern now is

$$E(x) = 1\ 1\ 0\ 0\ 0\ 0. \tag{6.13}$$

As this is not evenly divisible with $G(x)$, the error will be detected.

6.3. GENERATION OF THE CRC CHARACTER

As binary multiplication and division involves a series of add/subtract/shift operations, it is obvious that multiplier and divider networks can be constructed from shift registers and half-adder elements (exclusive-or circuits). A circuit for multiplication of any polynomial by the arbitrarily chosen $x^2 + x + 1$ polynomial is shown in Fig. 6.1. The boxes represent shift-register elements and circles half-adders. The multiplier was constructed according to the rules that the number of shift registers is equal to the degree of the polynomial. There is a half-adder between the shift registers if the subsequent term is one order lower,

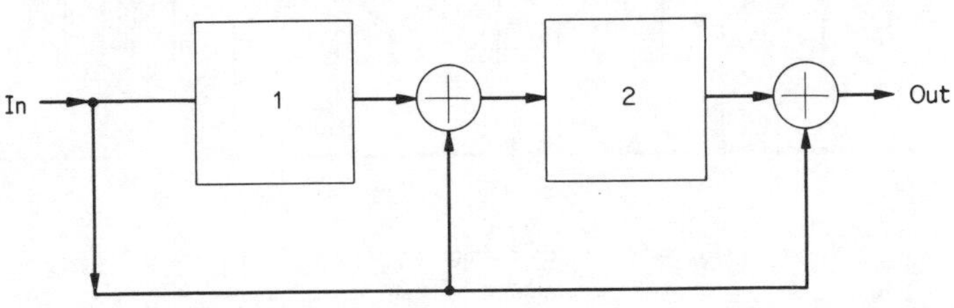

Shift No.	Register No. 1	Register No. 2	Output
Initial state	0	0	0
1	1	1	1
2	1	0	0
3	0	1	0
4	0	0	1

Fig. 6.1. Multiplication of $x + 1$ by $x^2 + x + 1$.

and no half-adder for terms with zero coefficient. The multiplicand is entered term-by-term at the input terminal and shifted through the registers. The total number of shifts is equal to the sum of the degrees of multiplier and multiplicand plus one. Figure 6.1 also shows the contents of the registers (which held initially zeros) at each subsequent shift as the arbitrarily chosen $x + 1$ multiplicand is multiplied (modulo-2) by $x^2 + x + 1$. For $1x^1 + 1x^0$ multiplier we entered 1, 1 subsequently, followed by two zeros. The result appears in the 'output' column as 1001. When checking the result by numerical multiplication the rules of modulo-2 addition should be remembered.

Two possible circuits – one parallel, the other for serial entry – for *division* by the generator function

$$G(x) = x^9 + x^6 + x^5 + x^4 + x^3 + 1, \qquad (6.14)$$

are shown in Fig. 6.2. This generator function was chosen for the 9-track NRZ1 format for encoding a block of data. The construction of the divider circuit follows some simple rules. The number of the shift-register elements is identical with the exponent of the highest power of x. Starting from the highest-order term, there is a half-adder between the shift registers if the subsequent term is one order lower. For terms of zero order there is no corresponding half-adder.

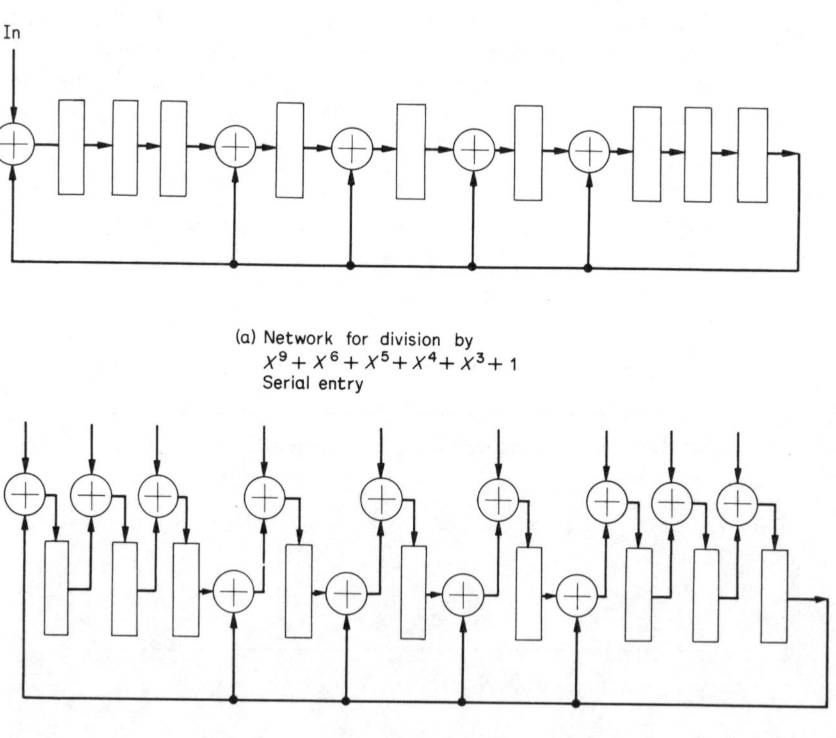

(a) Network for division by
$X^9 + X^6 + X^5 + X^4 + X^3 + 1$
Serial entry

(b) Network for division by
$X^9 + X^6 + X^5 + X^4 + X^3 + 1$
Parallel entry

Fig. 6.2. Networks for division by $x^9 + x^6 + x^5 + x^4 + x^3 + 1$.

In the series form of the divider circuit, the dividend is entered, highest-order term first, at the IN half-adder terminal; at the end of the division cycle the registers will contain the remainder of the division. The total number of shifts is one more than the exponent of the highest-order term of the dividend. In general, a block of data may contain m message characters. The message characters are entered into the register in parallel; each subsequent character – after shifting the previous character in the register by one position – is added via half adders. The net effect, as characters are shifted and added is *multiplication by the appropriate power of x and division of the sum by the generator function.* Thus the successive contents of the $G(x)$ register, as a block containing m message characters is entered will be as follows:

$$\text{Step 1} \quad \ldots \quad M_m$$

$$\text{Step 2} \quad \ldots \quad (M_{m-1} + xM_m) \bmod G(x)$$

$$\text{Step 3} \quad \ldots \quad (M_{m-2} + xM_{m-1} + x^2M_m) \bmod G(x)$$

The abbreviation mod $G(x)$ stands for the operation carried out by the network which corresponds to the $G(x)$ function on the characters entered. M_m is the first character in a block of m characters and M_1 is the last. The contents of the $G(x)$ register, after adding in the last character, will be

$$\sum_{i=1}^{n} (x^{i-1}M_i) \bmod G(x). \tag{6.15}$$

After entering the last data character one more all-zero character is added. The additional shift represents multiplication of x. If we write

$$\sum_{i=1}^{n} x^i M_i = M, \tag{6.16}$$

then the final contents of the $G(x)$ register can be written as

$$C_{\text{CRC}} = M \bmod G(x). \tag{6.17}$$

This C_{CRC} which has been generated by the method described is the Cyclic Redundancy Check character for the block of data.

It is recalled that the check bits of a systematic code were the remainder of the division of $M(x)x^{n-k}$ by the generator function $G(x)$. The CRC character is generated in the same manner but the encoding is carried out not on a single $M(x)_n$ character but on the

$$M = \sum_{i=1}^{n} x^i M_i, \tag{6.18}$$

block of characters.

6.4. EXAMPLE 3

The process of generating the check character will now be illustrated with an example. Let us assume that a block of data consists of the following five characters:

Character 1	0 1 0 0 0 0 0 1 1
Character 2	1 1 1 0 1 0 1 0 0
Character 3	0 1 0 0 1 0 1 0 0
Character 4	0 1 1 1 1 1 0 0 0
Character 5	0 1 1 1 0 1 0 1 0

The right-most bit is the least significant bit of the characters (LSB) and the left-most is the most significant bit (MSB). Each character consists of the 8 data plus one parity bit.

According to Equations (6.15) and (6.17) the CRC character is generated in the following manner:

$$
\begin{aligned}
\Sigma \,(\text{Mod } 2) \quad & [x^7 + x^1 + x^0] & . \, x^5 \\
& [x^8 + x^7 + x^6 + x^4 + x^2] \, . \, x^4 \\
& [x^7 + x^4 + x^2] & . \, x^3 \\
& [x^7 + x^6 + x^5 + x^4 + x^3] \, . \, x^2 \\
& [x^7 + x^6 + x^5 + x^3 + x^1] \, . \, x \\
\hline
& x^{11} + x^9 + x^8 + x^7 + x^5 + x^4 + x^2 .
\end{aligned}
$$

Dividing the sum by (6.14) will give the CRC character: (again it is the remainder of the division which is required)

$$
\begin{aligned}
&1\,0\,1\,1\,1\,0\,1\,1\,0\,1\,0\,0 \\
&1\,0\,0\,1\,1\,1\,1\,0\,0\,1 \\
\hline
&\quad\;\; 1\,0\,0\,1\,0\,1\,0\,0\,0\,0 \\
&\quad\;\; 1\,0\,0\,1\,1\,1\,1\,0\,0\,1 \\
\hline
&\quad\;\;\; 0\,0\,0\,1\,0\,1\,0\,0\,1 .
\end{aligned}
$$

Therefore, the CRC character for the above five data characters will be $x^5 + x^3 + 1$.

The logic circuit in Fig. 6.2 will give the same CRC character as that obtained by the mathematical operation.

The generation of the CRC character for a block of data can be explained from a different point of view. Let us take the first character from our example and generate the CRC for this single character. The operation

$$M \bmod G(x)$$

means that $x^7 + x + 1$ should be multiplied by x and divided by $G(x)$; the remainder of this division is the CRC_1 character or

$$M_1(x) \cdot x = (x^7 + x + 1) \cdot x = Q_1 G + R_1.$$

As now $Q_1 = 0$, $CRC_1 = R_1$ is generated by multiplying $x^7 + x + 1$ by x. The network, of course, gives the same answer. The generation of the CRC for the *first two* characters can be illustrated by

$$(R_1 + M_2)x = Q_2 G + R_2$$

where $CRC_2 = R_2$. The numerical operation gives

$$R_2 = CRC_2 = 1\,1\,0\,1\,0\,0\,1\,0\,0.$$

And for the first three characters,

$$(R_2 + M_3)x = Q_3 G + R_3$$

which yields, for $CRC_3 = R_3$, the value

$$0\,0\,0\,0\,0\,1\,1\,0\,0\,1.$$

The CRC for the four- and five-character block can be generated in the same manner. The CRC for the entire block is built up from the CRC's of the subsequent characters.

6.5. LOCATION OF FAULTY TRACK

During a writing operation the CRC character is generated for a block of data and recorded on the tape. The CRC is generated again from the information read back from the tape and the two CRC's are compared. If they are identical, the data block was read without errors.

If there is a difference between the write CRC and the read CRC, the error pattern is held in a register, which is of a construction similar to the CRC register (Fig. 6.3). Thus, for each shift, it multiplies by x and divides by $G(x)$. A logical 'one' is entered in the MSB position of the error pattern register (EPR) for each faulty character and shifted simultaneously with the CRC. This operation tests the error pattern whether it is divisible by $G(x)$.

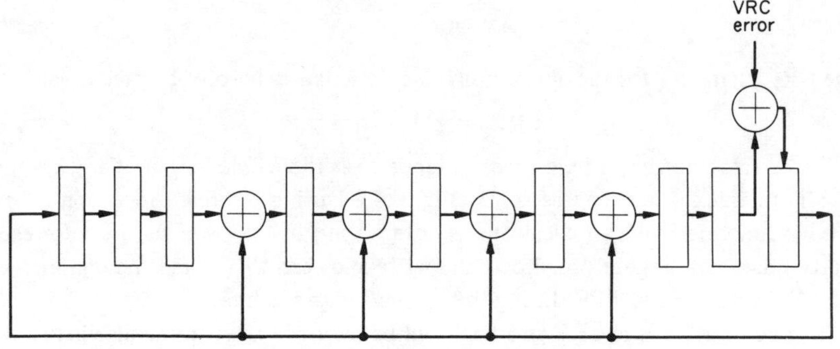

Fig. 6.3. Error pattern register.

The faulty character and the position of the error (i.e. the track position) can be located with the aid of the faulty character's own parity, (VRC), the CRC and the EPR register.

The EPR has one input to the MSB cell of its register. This input is a logical 'one' whenever an error occurs, otherwise it is zero. The EPR content is shifted simultaneously with the CRC, thus when the full block of data is read and the write and read CRC's compared, the correct bits will cancel out and a 'one' will remain in the position corresponding to the track in error. (Actual circuits are modified to give a non-zero 'match pattern'. This does not affect the operation of the error detecting properties of the system.) The CRC register's content is now shifted until it matches the content of the EPR; the number of shifts will give the track containing an error. If after eight shifts a match between CRC and EPR has not been obtained, then the error cannot be corrected.

The process is now illustrated in Example 4 (Fig. 6.4).

Track No.	Write CRC	Read CRC	Write CRC $\oplus$ Read CRC	Shift 1	Shift 2	Shift 3	Shift 4	EPR
0	1	1	0	0	0	1	1	0 1 1
1	0	0	0	0	0	0	1	0 0 1
2	0	0	0	0	0	0	0	0 0 0
3	1	1	0	0	0	1	1	0 1 1
4	1	1	0	0	0	1	0	0 1 0
5	0	1	1	0	0	1	0	0 1 0
6	0	1	1	1	0	1	0	0 1 0
7	0	0	0	1	1	0	1	0 0 1
8	0	0	0	0	1	1	0	1 1 0

Fig. 6.4.   Example 4. Register contents during error detection.

Consider a block of data consisting of the following three characters

$$x^7 + x^1 + x^0$$
$$x^8 + x^7 + x^6 + x^4 + x^2$$
$$x^7 + x^4 + x^2$$

The CRC character for the above block is calculated as before. The result is:

$$CRC_3 = x^4 + x^3 + x^0.$$

Let us assume that a read error extends over the whole length of track 4, i.e. all bits in track 4 are read as zeros (Fig. 6.4). During the read operation a new CRC is generated and a logical one is entered in Position 8 of the EPR for each faulty character. Figure 6.4 shows the write and read CRC's and the contents of the CRC Register and EPR during the correction procedure.

The contents of the CRC and EPR will be identical after four shifts hence the error is in track 4.

6.6. UNCORRECTABLE ERRORS

A number of read error conditions cannot be detected by the CRC read-error correction system. First of all, the error correction capability is limited to single track errors; any error extending to more than one track is uncorrectable.

Secondly, single track error patterns which are divisible by the generator function cannot be detected and corrected.

It should be noted that the chosen generator function is evenly divisible with $x + 1$:

$$G(x) = (x^8 + x^7 + x^6 + x^4 + x^2 + x + 1) . (x + 1) = G_0(x) . G_1(x).$$

The choice of this $G(x)$ was dictated, among others, by the consideration, that for error correction capabilities in backward reading mode, the generator function must be symmetrical and, for a polynomial with 9th degree, this requirement can only be satisfied if $G(x)$ has an $1 + x$ factor. For other considerations which dictate the choice of the generator function see references [1, 2, 3, 10].

If a correctable error is detected the program backspaces one record and reads the same again. The programming sequence for correcting a record is: read, test, backspace, correct, test, backspace, etc.

Quite often the error may extend over many bit lengths in one track. It can be shown [2] that a polynomial of $n - k$ degree will always detect a 'burst error' of $n - k$ length or less, but will also detect a high percentage of longer error burst as well.

6.7. ALTERATION OF THE CRC PARITY

For practical reasons it is desirable to have at least one bit in the LRC character which is not zero. To achieve this the final content of the CRC register is modified by the addition of the polynomial

$$G_2(x) = x^8 + x^7 + x^6 + x^4 + x + 1.$$

The modified CRC will now have a parity opposite to the unmodified CRC and this will cause the LRC parity to be odd, thus the LRC character has always at least one bit different from zero.

The choice of $G_2(x)$ polynomial for the purpose of altering the CRC character simplifies the procedure of locating the track in error. The consequence of adding $G_2(x)$ to the CRC register is that, on reading a correct record the final CRC content will not be all zeros, but $G_2(x)$. This $G_2(x)$ is often referred to as 'match pattern' in the technical literature.

If a block of data contains an error in track j, the final CRC register after addition of $G_2(x)$ will be

$$x^j E + G_2(x) \bmod G(x)$$

After shifting k times this will change to

$$x^{j+k}E + x^k G_2(x) \bmod G(x) = x^{j+k}E + G_2(x) \bmod G(x)$$

Thus the $G_2(x)$ term remains unaltered and can be subtracted simply by inverting the bit positions of the CRC corresponding to $G_2(x)$.

6.8. OTHER ERROR DETECTION METHODS

Many alternative error detection and correction schemes have been designed for non-interchangeable tape formats. When data reliability is more important than efficient utilization of the recording medium, redundant recording on two or three tracks can be employed; the information read back from three tracks is compared and if there is any discrepancy the 'majority decision' is taken as correct.

Error correction schemes have been implemented on low-speed, low-density data recorders by reading back the data just written and comparing it with the write data held in the write buffer. If there is any discrepancy, the control logic makes several attempts to re-write, and reports the error condition only if a specified number of attempts fail. Each time a re-write attempt is made, a check bit is written in one or two reserved tracks which do not carry data. On reading, the read control logic ignores those characters which have a check bit in the reserved track. Read error checking is based on parity detection.

A method frequently used on earlier designs was based on treating all characters in a block of data as binary numbers. The numbers were added and the 'check-sum' recorded at the end of the block. On reading the check-sum from the tape, it was compared with the check-sum for the block held in memory. An error checking method very often employed, consists of counting the number of characters in a block of data read from the tape, and compare it with the block length number held in the computer memory.

6.9. ERROR CHECKING ON NINE-TRACK PHASE ENCODED RECORDING FORMAT

The standardized 1600 bpi phase-encoded, recording format utilizes the vertical parity check character for error detection, but there is no longitudinal check character or cyclic redundancy character defined for this recording method. The logic in the magnetic tape controller would normally detect, if the identification burst, the preamble or postamble of a data block is not recognized. Various other checking facilities can be provided by the controller. As the phase-encoded format contains at least one flux transition for each data bit, the loss of data caused by a drop-out can be easily detected. Typically this error detecting circuit will contain two bistables, one of which is set by a logic zero, the other by a logic one. If neither of them are set during the data period, an error condition is indicated. Single bit errors in one track can be reconstructed with the aid of the parity check bit of the character in which the error occurred. The read electronics can be made tolerant towards fluctuations of readback amplitudes, and a readback level as low as 15% of the normal can still be accepted as a valid data signal.*

The read circuits can include an 'envelope check', i.e. the envelope of the readback signal is monitored and the read amplifier is enabled only if the envelope signal is present. The errors which follow each other immediately in a single track, or in several tracks, are counted, and if the count reaches a pre-set number, an error signal is generated inhibiting the track in error ('dead tracking') as its phase-locked oscillator which is normally re-synchronized at every data bit may lose synchronism.

REFERENCES

1. Richards, J. K. (1971), *Digital design,* Wiley-Interscience, Chapter 5.
2. Peterson, W. W. and Brown, D. T. (1961), 'Cyclic codes for error correction'. *Proc. IRE,* **49,** 228–235.
3. Peterson, W. W. (1968), *Error correcting codes,* MIT Press.
4. Brown, D. T. and Sellers, F. F. (1970), 'Error correction for IBM 800 bit per inch magnetic tape', *IBM J. of Res. and Dev.,* **14,** 4, 384–389.
5. Massey, J. L. (1963), 'Error correcting codes applied to computer technology', *Proc. National Electronic Conference.*
6. Lin, Shu (1970), *An Introduction to Error Correcting Codes,* Prentice Hall.
7. Liccardo, M. A. (1971), 'Polynomial error correcting codes and their implementation', *Computer Design,* **10,** 8, 53–59.
8. Wallner, A. (1972), 'Error detection for peripheral storage devices', *Computer Design,* **11,** 1, 57–61.
9. Sellers, H. F., Hsiao, M. Y. and Bearnson, L. W. (1968), *Error Detecting Logic for Digital Computers,* McGraw-Hill, 251–269.
10. Tang, D. T. and Chien, R. T. (1969), 'Coding for error control', *IBM Syst. J.,* **8,** 1, 48–86.

* Some controller designs claim readability at 5% of nominal level.

7 Tape handling mechanisms

7.1. DIGITAL TAPE TRANSPORTS

The primary feature of digital tape mechanisms – also named tape transports, tape handlers – which distinguishes them from the audio and instrumentation recorders is their near-instantaneous start-stop-reverse capability under computer control. The operation of the computer demands that only a small amount of data is transferred at one time to or from the computer, hence in a typical operation mode the tape transport is requested to run for a short period then stop and wait for the computer to request or transfer further data. The information is recorded in blocks with unrecorded sections – interblock gaps – between them. Efficient utilization of the tape requires that the interblock gaps, which are wasted as far as data storage is concerned, should be as short as possible, which in turn necessitates fast starting and stopping of the tape. If the tape is stopped after reading a block of data and then re-started, it must again reach full stable velocity at the beginning of the next block otherwise read errors can occur. The total start-stop (or reverse) distance – the length of tape moved during acceleration and deceleration – must be less than the interblock gap length. Typically, start time, from issue of the command to 90% of the final velocity is between 3 and 10 ms for high-speed high performance transports and below 20 ms for low-speed, low-cost devices. The stop time is usually the same as start time. The length of the interblock gap is standardized for inter-changeable tapes at 0.5 and 0.75 in nominal, but some standards permit a maximum length of 25 ft.

In addition to the forward and reverse mode of operation digital transports normally provide rewind/spool facility at a speed considerably higher than the data transfer speed. During the high speed rewind/spool operation the head is retracted or the tape is lifted off the head. Some transports designs permit high-speed 'search' operations which save time in finding a particular block of data.

Tape velocities in the read-write mode of operation range from 10 ips at the low-speed end to exceptionally 300 ips but few commercial designs utilize more than 200 ips. Typical low-speed transports operate at 10, 16, 24, $37\frac{1}{2}$ ips speed, medium-speed devices at 45 to 80 ips and high-speed transports at $112\frac{1}{2}$, 150, 200 ips and above.

The electro-mechanical design of the tape transport must satisfy a number of sometimes conflicting requirements in handling the delicate and vulnerable magnetic tape. Starting, stopping, and the reversing operation must not stress the tape to an extent which might damage data integrity; the oxide coated side of the tape should not be in contact with any other member of the transport but the heads, and even with those only at write/read/search operation. The design of the tape path and tape guides should ensure that dynamic skew is minimal. A correct winding tension pattern is essential for safe storage of the tape on the reels. The tape speed must be constant to a high degree; typically, long-term variation not more than 1% is necessary. Interlocks on the machine should prevent any sequence of operator intervention which may cause damage to the tape, or accidentally erase data. Loading the tape onto the transport should be simple, preferably fully automatic. Finally the requirement of interchangeability of tapes between machines imposes very close tolerances on all mechanical and electronic components of the transport.

In this chapter we will review the operation and main functional elements of digital tape transports. The design of such machines is a discipline of servo-control, precision mechanical and electrical engineering and as such, it is beyond the scope of this book. Some elements of digital tape transports are specified in the relevant standards.

7.2. TAPE TENSIONING AND BUFFERING

The design of digital tape transports is governed by the necessity of repetitively starting and stopping the tape within a few milliseconds. It is obvious that to provide the drive power for accelerating heavy reels of tape in such short time is not a practical solution; instead a short length of tape in the vicinity of the heads is isolated by buffers on both sides of the head, from the feed-out and take-up reels. It is only this short length of tape which needs to be accelerated and declerated; the buffers isolate it from the reels and hold sufficient tape to allow the reels a reasonable time to reach the operating speed, or to stop. The amount of tape in the buffers is monitored and the reels are servo-controlled so that they tend to keep the length of the loop constant. At the same time the tape tension is kept at a constant value, independently of the amount of tape on the supply and take-up reel.

The principles of two basic tape buffering arrangements are shown in Figs. 7.1 and 7.2. On low-speed transports the isolation of the reels is provided by spring-loaded, often servo-controlled tension arms which keep a loop of tape ready to be supplied to the reel or to the tape drive capstan. Sensing the position of the tension arm provides the feedback to the reel drive servo which feeds out or takes up tape so that the servo arm remains in the centre of its range of travel. This arrangement is simple and inexpensive but the inertia of the driven length of the tape is increased by some of the tension arm inertia. The tension arm and roller causes somewhat rough tape handling and not only needs careful adjustment, but imposes an upper speed limit – usually around 40 ips – where

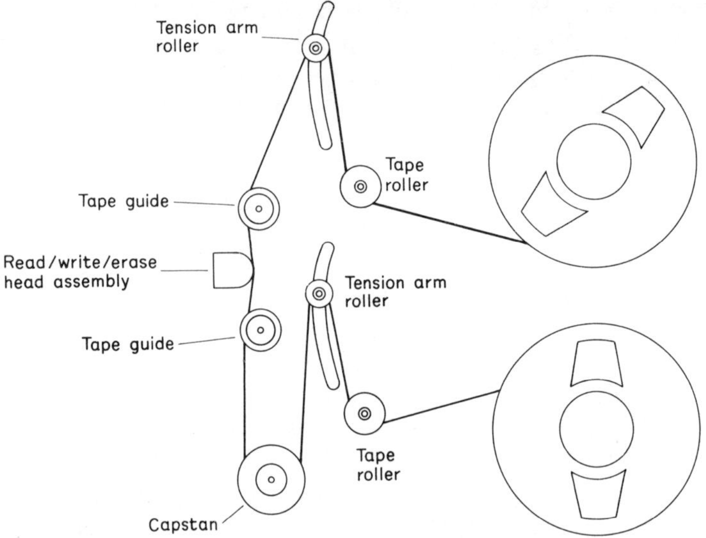

Fig. 7.1. Schematic of a low-speed tape drive with tension arm tape buffer and single-capstan drive.

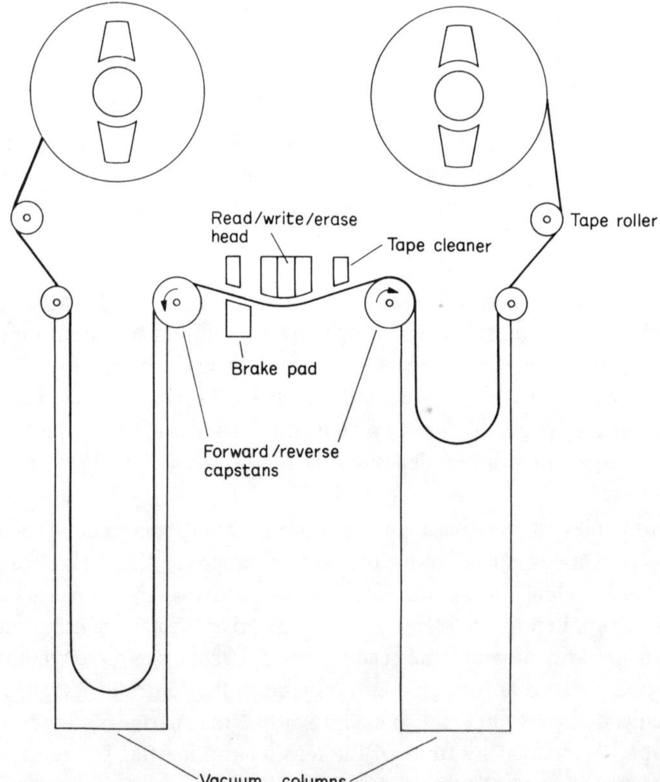

Fig. 7.2. High-speed vacuum-column buffer dual capstan tape transport.

the use of the tensioning arm becomes impractical. The tensioning arms can be servo-controlled with multiple tape loops so that the length of tape in the loop increases, permitting higher tape speeds, but the design looses its simplicity.

Medium- and high-speed high-performance tape transports use a vacuum system for buffering and defining the tape tension. The tape is pulled into twin columns, placed between the tape drive mechanism and reels, by the reduced air pressure beneath the tape loop. Tape-position sensing ports, operating on vacuum-sensing, photoelectric or magnetic principle, monitor the position of the tape in the column and provide feedback for the reel motors and vacuum pump. In a simple system only on-off servos are used instead of proportional control. As the tape handler moves the tape at constant speed across the heads, the reel motors operate in a start-stop mode. Figure 7.2 shows the tape paths of a typical dual-capstan vacuum buffer high-speed transport. During forward operation the forward capstan (see also paragraph 7.3) pulls the tape from the reverse column and places it into the forward column. The reverse reel servo replaces the tape in the reverse column. As the tape is placed in the forward column the slack in the tape loop is taken up by the forward reel servo.

Adjacent to the read/write and erase head, two tape cleaners are shown. They are miniature suction cleaners which remove minute particles – dirt, dust, loose oxide – from the recording surface. The reel and capstan motors have electro-mechanical brakes. Fast stopping is enhanced by a vacuum tape brake consisting of a hollow chamber, pressure valve and solenoid; the tape is held against the perforated brake surface when the vacuum is applied. Positive pressure during tape motion causes the tape to float above the brake surface.

The operator's task of loading the tape on the transport is simplified, on some high-performance handlers, by a fully automatic loading system. Once the tape, which is stored in a cartridge, is placed in position on the transport, the servo and vacuum systems wind the tape onto the take-up reel and place an appropriate length of the tape in the buffers. Accidental erasure of tape is prevented by a plunger type switch which inhibits writing when the contacts are open. The switch is closed by a plastic ring which fits into a groove moulded into the reel (Fig. 7.3). When the reel is mounted on the supply hub, the ring closes the plunger switch and enables the write circuits. After writing is completed, the protection ring is removed.

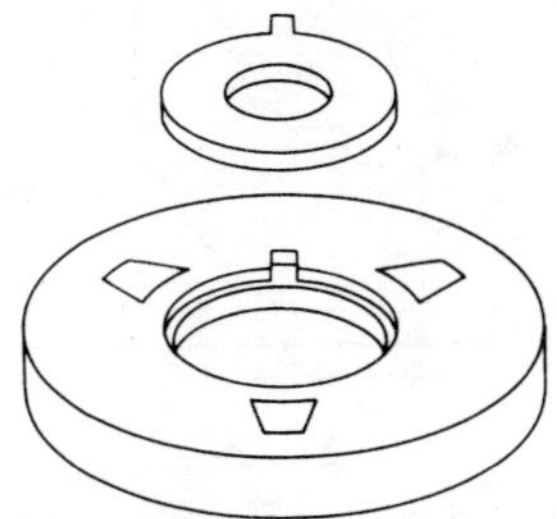

Fig. 7.3. Write protection ring.

7.3. TAPE DRIVE MECHANISMS

The purpose of the tape drive mechanism is to move the tape across the head at constant velocity and to start, stop, and reverse tape motion within a few milliseconds. The basic mechanism consists of a rotating cylinder (capstan) and a roller with an axis parallel to that of the capstan (pinchroller). The tape motion is caused by the pinchroller pressing the tape against the rotating capstan; the tape velocity will be, as long as there is no slip between the tape and capstan, the same as the peripheral velocity of the capstan. Continuous rotation of the capstan avoids the need of rapid acceleration and deceleration on start/stop commands. The pinchroller movement, which is accomplished by a solenoid or voice-coil, is in the same order of magnitude as the thickness of the tape. The capstan-pinchroller drive is extensively used on audio and instrumentation transports but on digital tape handlers it is superseded by the *vacuum capstan* [1, 2] and the *single-capstan* [3] drives, which permit shorter start-stop times, better tape handling, and longer trouble-free operation.

A vacuum capstan is basically a hollow cylinder with tightly spaced narrow axial air slots on its periphery. A vacuum-pump and a positive-pressure pump are connected to the cylinder. The tape is engaged by opening a valve which connects the capstan to the vacuum source and stopped by applying positive pressure to the capstan. The stop time can be reduced by applying vacuum as a braking force to the tape directly through a slotted 'vacuum brake'. In tape drive mechanisms, using two vacuum capstans, the two capstans rotate continuously in opposite direction; tape motion is accomplished by applying vacuum to one and positive pressure to the other.

The air-pressure change must be rapid, so the valves, which control the pressure change are located close to the slots in the capstan minimizing the lag caused in the air passages. The vacuum capstan gives reproducible and short start-stop times and handles the tape more gently than the pinchroller drives.

The principle of single capstan drive is illustrated in Fig. 7.4. The tape is held

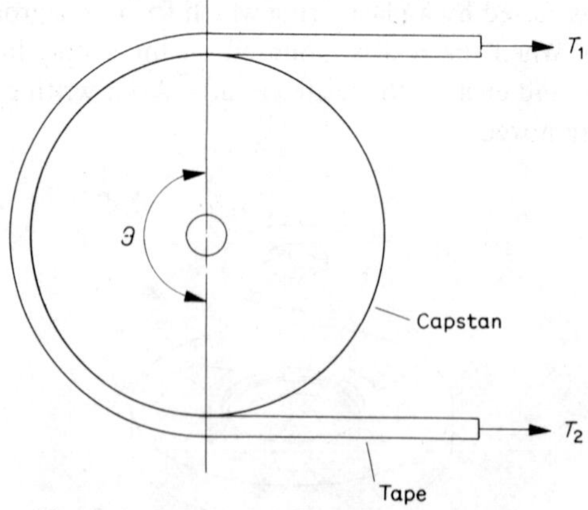

Fig. 7.4. Principle of single-capstan drive.

in contact with the capstan at all times by the constant tape tension derived from the vacuum-column or tensioning arm servo. The tape motion follows the capstan surface motion provided that the friction between tape and capstan is above a minimum.

The relation between tensions T_1, and T_2 is given by the well-known law:

$$\frac{T_1}{T_2} = e^{\mu\Theta}. \tag{7.1}$$

Here μ is the friction coefficient between tape and capstan surface and Θ is the wrap angle over the capstan. For a given value of T_2 tape tension, the maximum driving force, F_{max} which can be developed before slipping occurs, is

$$F_{max} = T_1 - T_2 = T_2(e^{\mu\Theta} - 1). \tag{7.2}$$

As start and stop times in this drive system are determined by the acceleration and decleration of the capstan (including the clutch or motor inertia if there is no clutch between capstan and drive motor shaft) the latter is made of light alloy with its tape driving surface coated by a suitable material* providing high friction between the tape and capstan. Low-inertia printed-circuit motors permit

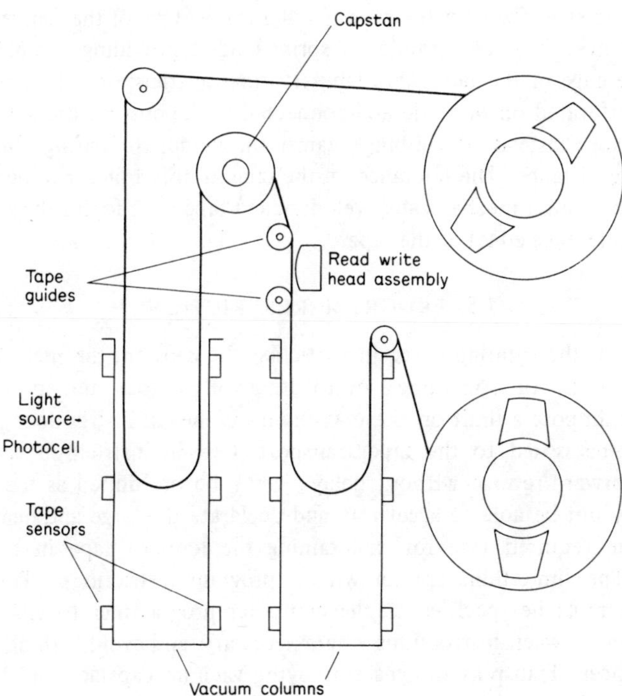

Fig. 7.5. Medium-spaced single-capstan tape drive with vacuum-column buffers.

* Adiprene, Neoprene, Polyurethane.

the mounting of the capstan directly onto the motor shaft and still achieve a short, start-stop time and distance; this makes the design simple and economical.

A typical tape path arrangement, utilizing a single-capstan drive at medium speed, is illustrated in Fig. 7.5. The position of the tape in the columns is monitored by a light source-photocell arrangement. The beginning of tape/end of tape reflective markers are recognized by another light source and photocell. If the tape in the loop hits the bottom of the vacuum column, an 'excessive slack detector' corrects the servo operation or initiates a systematic shut-down procedure, preventing tape damage.

7.4. TAPE GUIDANCE

The accuracy of the tape guidance is one of the factors which determines the maximum practical recording density in parallel recording systems. Unwanted lateral motion of the tape – which can be caused by a minute misalignment between capstans, head and guides – is the main source of time delays between the two outer tracks, and sets a limit on bit packing density achievable without complex electronic de-skewing. The tape guides most frequently used on digital transports, are of the 'fixed' type, i.e. a non-rotating member over which the tape passes. A typical guide of this type is constructed of hard, highly polished non-magnetic steel fixed to the tape handler frame. One of the flanges, made of hard substance, such as ceramic, is spring-loaded providing a small pressure against one edge of the tape. On high-performance transports the guide may be tubular, perforated on one side and connected to the positive pressure source so that the tape, instead of rubbing against the guide, is floating above it at a microscopic distance. The tolerance on the tape itself is a factor to be taken into consideration and a mechanically well designed tape path forces the tape against one edge (reference edge) of the tape.

7.5. PROGRAM RESTRICTIONS

The inertia of the rotating parts, in particular the capstan, the mechanical tape tensioning arms, or sometimes overheating of motors and power supply restrictions, impose a limit on the maximum number of instructions per second which can be issued to the tape transport. Certain instruction sequences – typically forward/reverse without delay – may be prohibited as the reel servo system may not be able to accelerate and declerate the large and heavy reels of tape at the required rate for maintaining the loop of tape in the vacuum columns. The limitations are known as 'program restrictions'. The program restrictions must be specified to the computer programmer to enable him to insert delays between instructions where necessary, and avoid forbidden instruction sequences. Transport designs employing vacuum capstans and buffers are usually free from programming restrictions; other designs which use servo-controlled mechanical tape tensioning arms often have a speed limit above which program restrictions are specified.

7.6. CASSETTE AND CARTRIDGE RECORDERS

There are many applications for magnetic tape as a data storage medium where there is no need for the capacity and transfer rate which the large parallel recorders can provide. Communication terminals can be made more flexible by addition of a small tape memory; processors using volatile semiconductor memories need a non-volatile backing store. Punched cards and punched paper tape, which are the basic program preparation mediums for a wide range of computers, can be replaced by magnetic tape requiring much smaller space for the same data storage capacity. The need for a low-cost small-capacity magnetic storage system led to the development of a wide range of series-by-bit, series-by-character recorders employing narrow tape in a protective container. It is customary to distinguish between 'cassette' recorders in which the tape can be driven in either direction from reel-to-reel without removing the tape from the cassette, and 'cartridge' recorders which permit unidirectional tape motion only and contain an endless loop of tape.

Early attempts to use the ubiquitous audio cassettes and cartridges for digital data recording were largely unsuccessful. The reliability and error rates of audio recorders were unacceptable for data recording. Only since a new generation of cassette recorders now specifically designed as digital memories came on the market have cassette recorders gained acceptance as computer peripherals. Tape motion can be acccomplished by the conventional capstan/pinchroller drive system and tape tension is controlled by applying torque to the reels by separate reel motors or slip clutches. The cassette or cartridge construction does not permit the use of tape loops, isolating the reels from the drive so the length of tape and tape speed are limited to low values. Precision drives have the capstan motor fitted with tacho generator to provide feedback to the speed-control servo. Such cassette recorders can have a read-after write, bidirectional read and search facility. Some systems provide a pre-recorded address track making possible access and read or write data in the selected sector without affecting others.

A wide range of recording techniques and formats are used. Many systems use a second track to enhance reliability. The second track may carry the complement of data making the NRZ1 coding self-clocking (Fig. 7.6). Alternatively data are split between the two tracks in such a manner that one carries the logic zeros the other the logic ones.

A dual track recording system which is insensitive to tape speed variation at the expense of poor utilization of the tape is illustrated in Fig. 7.7. The use of two tracks permits definition of four different states of magnetization. In State 1 both tracks are saturated in the same (negative) direction; this is the condition not only in the interblock area but also between bits and characters. A logic 1 is represented by a positive saturation pulse on track 1, while track 2 remains unchanged and a logic zero is recorded as a positive saturation pulse on track 2 while track 1 is unchanged. The fourth condition, when both track 1 and track 2 change to positive saturation, is used as an end-of-byte signal. The zero and one

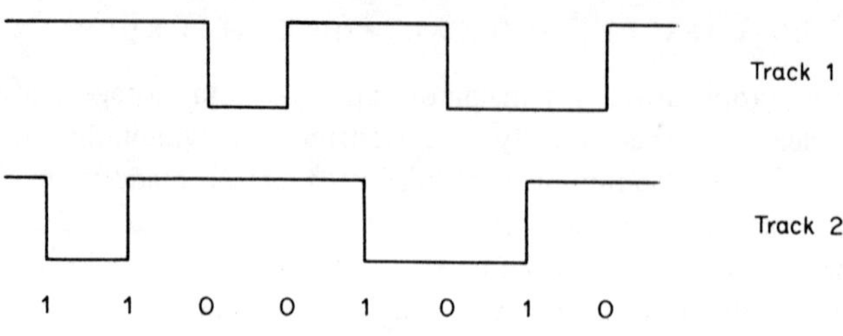

Fig. 7.6. Dual track NRZ1 coding.

bits are separated from each other by the erased area; in Fig. 7.7 the length of the erased area is shown to be equal to the pulse lengths which represent data, but other convenient values can be chosen. The system is self-clocking and has

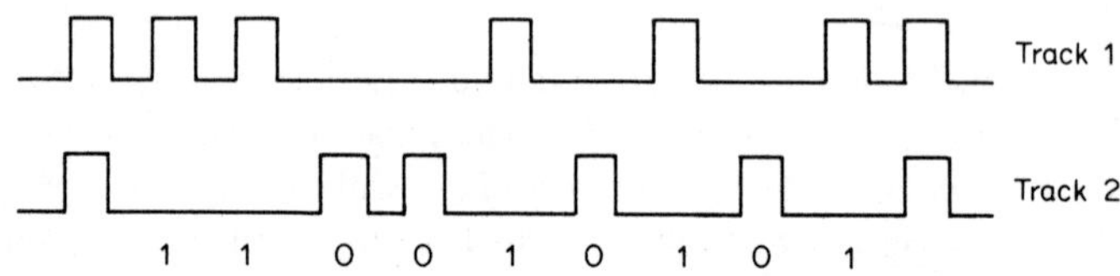

Fig. 7.7. Dual track four-phase coding.

the feature that each data byte is delimited by the end-of-byte signal. This permits the use of an error-detection which in its simplest form consists of a three-bit counter; any dropout is detected as erroneous count. The character in error can be easily detected by some additional logic. As there is no need for synchronization as in PE coding, detection circuits are very simple. Again, this system has many variants.

The standards which have started emerging for cassette tape are based on the widely used Philips audio cassette with 0.15 in (3.81 mm) wide tape (ECMA-34, 1971, Fig. 7.8) or use a wider cassette for $\frac{1}{4}$ in wide tape.

The ECMA-34 standard specifies 800 bits/in (32 bit/mm) phase encoded recording on two tracks. The length of tape is 282 ft (86 m). The cassette which has approximately 4 in by $2\frac{1}{2}$ in main dimensions contains two flangeless reels. The reel hubs are engaged by the drive shafts of the tape transport. The friction torque of full hubs, a pressure pad to provide adequate head-to-tape contact, all dimensional and operational characteristic of cassette and tape are carefully specified in order to allow full interchangeability. A cyclic redundancy character is proposed for error correction.*

* The chosen generator function

$$G = x^{16} + x^{15} + x^2 + 1$$

permits detection of burst errors of 16 bits or shorter; 99.985% of longer bursts in a block equal or less than 4096 bytes length.

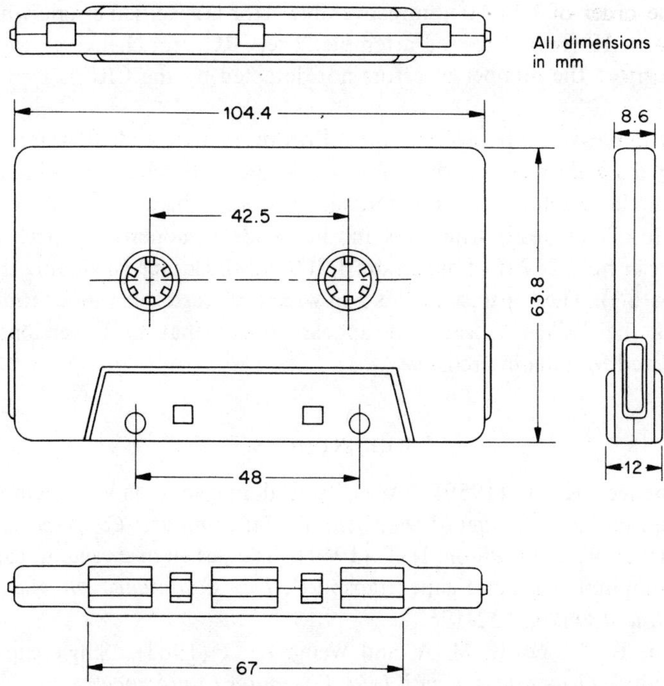

Fig. 7.8. ECMA cassette for 3.81 mm tape.

The thickness of the tape specified in the ECMA standard is 19 μm including the 5 μm thick coating. The storage capacity as function of the block length (bytes per block) for a cassette recorder conforming to ECMA-34 is shown in Fig. 7.9. The eights bit in each byte must be zero. Recoverable read error rates

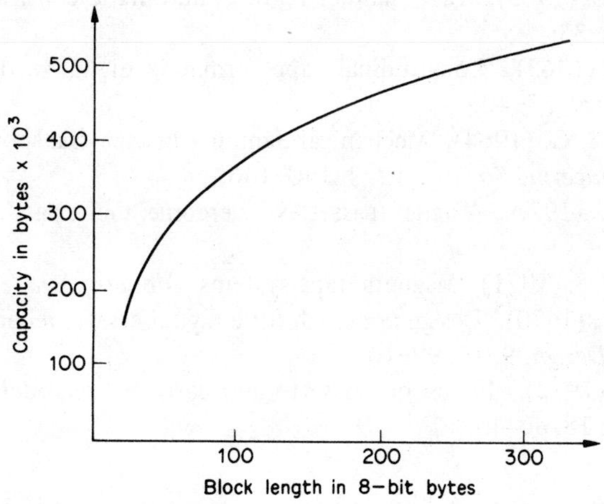

Fig. 7.9. Data storage capacity of an ECMA cassette as function of block length.

are in the order of 1 in 10^7; unrecoverable error is specified as one which is not corrected in 10 (or 15) re-read attempts. The CRC error checking provides high data integrity: the number of errors not detected by the CRC is typically 1 in 10^{10}.

The alternative proposal for standardization recommends 6.30 mm (approximately $\frac{1}{4}$ in) wide tape on which four tracks are recorded at 1600 bit/in density in PE mode. Each track is recorded bit-serial, character-serial so that the alignment of the read/write gaps in the heads is noncritical. The $\frac{1}{4}$ in tape cassette can hold 282 ft of maximum 1.200 mil thick tape (including the 0.2 mil thick coating). The capacity of this $\frac{1}{4}$ in wide tape cassette is about four times as much as the ECMA cassette; it appears likely that both versions will be standardized for data interchange.

REFERENCES

1. Lawrence, R. B. (1959), 'An advanced magnetic tape system for data processing', *Proceedings of the Eastern Joint Computer Conference.*
2. Willis, D. W. and Gwillim, D. T. (1964), 'The design of a vacuum capstan for a computer magnetic tape transport', *Int. Conference on Magnetic Recording, London,* 122–125.
3. Kleist, R. A., Lewis, M. A. and Wang, B. C. (1963), 'Single capstan tape memory', *Proceedings of Fall Joint Computer Conference.*
4. Davies, G. L. (1961), *Magnetic tape instrumentation,* McGraw-Hill, 75–96, 247–253.
5. Harrison, H. M. (1960), 'The development of a high performance tape handler', *J. Brit IRE,* **20,** 841–850.
6. Franklin, D. P. (1960), 'Factors influencing the application of magnetic tape recording to digital computers', *J. Brit. IRE,* **20,** 9–21.
7. Kienle, J. E. (1963), 'Tape handler features automatic control', *Automatic control,* 30–33.
8. Gabor, A. (1963), 'Longitudinal tape format in digital recording', *Automatic Control,* 35–40.
9. McKnight, J. G. (1964), 'Mechanical damping in tape transports', *J. of the Audio Engineering Society,* **12,** 2, 140–146.
10. Hovey, M. (1970), 'Digital cassettes overcome their past', *Electronics,* 129–134.
11. Murphy, W. J. (1971), 'Magnetic tape systems', *Modern Data,* 32–45.
12. Sykes, J. R. (1970), 'Design approach for a digital casette recording system', *Computer Design,* **9,** 10, 99–103.
13. Kaye, N. (1972), 'Focus on cassette and cartridge recorders', *Electronic Design,* **20,** 18, 40–46.

8 Future trends

It has been recognized since the dawn of history that predicting the future is sometimes dangerous and nearly always unrewarding. The history of science abounds with examples when the foremost and unchallenged authorities of the day declared an idea impossible, and to be in conflict with the Laws of Nature, only to find it implemented a relatively short time later.

Only after the Second World War did digital computers leave the scientific laboratories and become part and parcel of everyday life. As many computers serve primarily as data banks, the need for large capacity inexpensive memories, which provide rapid access, occupy small space, and are, for practical purposes, error-free, has channelled vast research efforts in the field of man-made memories. Whether any new technology will emerge which will satisfy simultaneously all requirements, and thus render magnetic surface storage obsolete, is a matter of speculation. The major breakthrough with laser/optical memories, which had been predicted for a decade, did not materialize. On the other hand, semiconductor memories are incorporated in practically all of the latest generation of computers, and appear to be replacing the traditional core mainframe memories. Volatile semiconductor memories need a non-volatile back-up and it appears that magnetic surface stores are eminently suitable for this role.

We believe that the principal factor which will decide in favour of magnetic surface storage is the capability of retaining recorded data indefinitely without electrical energy, and at the same time, easy erasability. On the debit side is the need for a complex precision mechanism, in total contrast to the philosophy of the central processor memory which has no moving parts.

The maximum track density is dictated by a number of considerations which we have reviewed in the relevant chapters: the problems of manufacturing multi-track in-line heads, signal-to-noise ratio limitations, tape slitting tolerances, etc. It is unlikely that there will be a dramatic increase in area density for data interchange type of applications. This does not apply to discs. On discs a surface density of $1.5 \cdot 10^6$ bits per square in can be achieved with current techniques and as novel integrated head construction are very suitable for disc-type applications it is not unreasonable to expect a surface density of $5 \cdot 10^6$ per square in in the next five years.

We have divided the field of digital magnetic tape recording into such categories as recording materials, heads, recording techniques and transports. It seems that the greatest potential for improvement in density and reliability is in the areas of recording media and heads. Integrated heads with very thin high-coercivity recording media seem to point in the direction in which substantial improvements can be achieved. The next logical step appears to be to combine integrated heads with integrated-circuit electronics using same technology for both; combined with large-scale integration and firmware approach to the controller logic a drastic decrease in size of the tape memory unit is likely to be achieved.

The requirement that heads and tapes shall be interchangeable imposes severe constraints on the tape transports. Only slow improvements are expected in the tape handling capability of the conventional transports, such as higher forward/rewind speed and better tape handling. Little use has yet been made of the techniques developed for video recording in applications for data storage. In such machines the track is orthogonal to the tape motion and the revolving head provides high head-to-tape speeds. Track width of 3.5 mil is currently used and 250 tracks in density seems possible for the future. It appears to be somewhat unlikely that conventional reel-to-reel transports will be able to offer much improved access time; instead access time and transfer rate will be improved by multiple data channels.

The progress of magnetic surface recording has been somewhat slow and unspectacular over the last two decades. The 200 bits/in density of the 1960's has increased to 1600 bits/in for interchangeable format and 6250 bits/in is expected to become the industry standard of the future. The error rates improved by some three or four orders of magnitude. Our forecast is that magnetic surface recording will play an important role in data storage for at least one, possibly two decades in the future.

9 Magnetic units

The student of applied electricity and magnetism may encounter as many as four different systems of units in his textbooks such as the electrostatic system (ESU), the electromagnetic system (EMU), the rationalized and nonrationalized MKS system and the Gaussian (sometimes called mixed CGS) system of units.

The ESU and EMU systems with the fourth unit of charge and current, respectively, are sometimes referred to as the CGS-Franklin and CGS-Biot where

Quantity	Symbol for quantity	MKSA (SI) unit	CGS (practical) unit	Conversion factor CGS/SI*
Magnetic field strength	H	ampere/metre A/m	oersted Oe	$\dfrac{10^3}{4\pi}$
Magnetic induction	B	tesla T	gauss G	10^{-4}
Magnetic flux	ϕ	weber Wb	maxwell Mx	10^{-8}
Magnetization†	M	ampere/metre A/m	oersted Oe	10^3
Intensity of magnetization (magnetic polarization)‡	J	tesla T	gauss G	$4\pi.10^{-4}$
Magnetic constant	μ_0	henry/metre H/m	gauss/oersted G/Oe	$\dfrac{4\pi}{10^7}$
Magnetomotive force	F	ampere A	gilbert Gb	$\dfrac{10}{4\pi}$
Reluctance	R	1/henry 1/H	gilbert/maxwell	$\dfrac{10^9}{4\pi}$
Magnetic moment (area)	m	ampere/metre^2 A.m^2	— A cm^2	10^{-3}
Magnetic diple moment	j	weber/metre Wb.m	maxwell cm Mx.cm	$\dfrac{4\pi}{10^{10}}$

* Factor by which a quantity expressed in CGS units is multiplied to obtain it in SI units.
† Defining equation: $M = (B/\mu_0) - H$.
‡ Defining equation $J = B - \mu_0 H$.

Fig. 9.1. Magnetic units.

the Franklin (Fr) is the unit of charge and Biot (Bi) the unit of current. To distinguish between the units of charge and current in ESU and EMU the names statcoulomb (= franklin), statampere and abcoulomb, abampere (= biot) are also used. The CGS systems are non-rationalized.

The MKS system with the ampere as the fourth unit is known as MKSA. The rationalized MKSA units are part of the SI system of units (Système International d'Unites) which has six basic units (meter, kilogram, second, ampere, kelvin, candela).

The CGS Practical System is based on the EMU with the ampere as the unit for current and addition of volt, ohm, joule, watt for electric potential (EMF), resistance, energy and power, respectively.

Although the rationalized MKS system has been widely adopted in the field of applied sciences and engineering, textbooks and even recent papers on electricity and magnetism often use Practical CGS for some and MKSA units for other quantities, as convenient.

Figure 9.1 shows the numerical relations and distinctive names for the most important magnetic quantities in Practical CGS and MKSA (SI) systems. The terminology follows [1].

9.2. BASIC RELATIONS

In free space

$$B = \mu_0 H \tag{9.1}$$

where B is the magnetic induction (magnetic flux density), H the magnetic field strength (magnetic field, magnetizing force) and μ_0 the magnetic constant (permeability of free space). The SI units of the quantities in Equation (9.1) are:

$$[B] = \frac{V_{sec}}{m^2} = \frac{Wb}{m^2} = T \qquad \text{(tesla)} \tag{9.2}$$

$$[\mu_0] = \frac{V_{sec}}{Am} = \frac{H}{m} \qquad \text{(henry/metre)} \tag{9.3}$$

$$[H] = \frac{A}{m} \qquad \text{(amper/metre)} \tag{9.4}$$

and the numerical value of μ_0 is:

$$\mu_0 = 4\pi 10^7. \tag{9.5}$$

In presence of magnetizable material

$$B = \mu_0 H + J = \mu_0(H + M). \tag{9.6}$$

In Equation (9.6) the symbol J stands for magnetic polarization (intensity of magnetization, magnetization, intrinsic flux density) which expresses the increase of flux density over the flux density that would exist in vacuum with the same magnetic field strength.

The relative permeability μ_r of a material is the ratio of the magnetic flux density produced in it to that produced in a vacuum by the same magnetizing force; the absolute permeability μ of a material is equal $\mu_0\mu_r$. In engineering practice 'permeability' – unless specified otherwise – means the relative permeability of the material.

The quantity M in Equation 9.6 is the magnetization of the material (a term sometimes used for J or B_i). It should be noted that J is defined by

$$J = B - \mu_0 H, \tag{9.7}$$

whereas M is defined by

$$M = \frac{B}{\mu_0} - H, \tag{9.8}$$

and that the dimension of J is the same as that of B; the dimension of M is equal to that of H.

We have

$$B = \mu_0 H + J, \tag{9.9}$$

and

$$B = \mu H = \mu_0\mu_r H, \tag{9.10}$$

therefore

$$J = \mu_0\mu_r H - \mu_0 H = \mu_0 H(\mu_r - 1) = \mu_0 H\kappa \tag{9.11}$$

and

$$B = \mu_0 H(1 + \kappa). \tag{9.12}$$

The term $\mu_r - 1 = \kappa$ is the magnetic (volume) susceptibility. Since it is easier to measure mass than volume, the quantity termed mass susceptibility is often introduced:

$$K = \frac{\kappa}{\rho}, \tag{9.13}$$

where ρ is density. The mass susceptibility has the dimension of m^3/kg in SI.

In CGS system of units the Equation (9.6) takes the form:

$$B = H + 4\pi J. \tag{9.14}$$

In electromagnetism it is necessary to distinguish between two moments: that arising from the magnetic effect of a current loop and that of a magnetic dipole. The first is termed as the electromagnetic moment (or magnetic moment) m, the second as dipole moment or magnetic moment j. In Si system the units are ampere metre^2 and weber metre, respectively.

The intensity of magnetization, J, can be defined as the magnetic (dipole) moment per unit volume:

$$J = \frac{j}{V} \tag{9.15}$$

which naturally yields the same dimension, V sec/m^2 or Wbm/m^3 for J as Equation (9.7).

Some textbooks and papers on magnetic recording use a different terminology and symbols in particular by denoting the magnetic moment per unit volume by M. Others use M for magnetic dipole moment. In our Chapter 4 equations are quoted from the original papers in unaltered form; it can be established from the context which quantity, $B - \mu_0 H$ or $B/\mu_0 - H$ is meant.

9.3. THE HYSTERESIS LOOP

The induction B plotted against the magnetizing force H for a magnetic material, exhibits the characteristic shape shown in Fig. 9.2. The value of H beyond which there is no further increase in B, the saturation point, is marked with B_s on the curve. Reducing the magnetizing force from H_s to zero, the remanence or

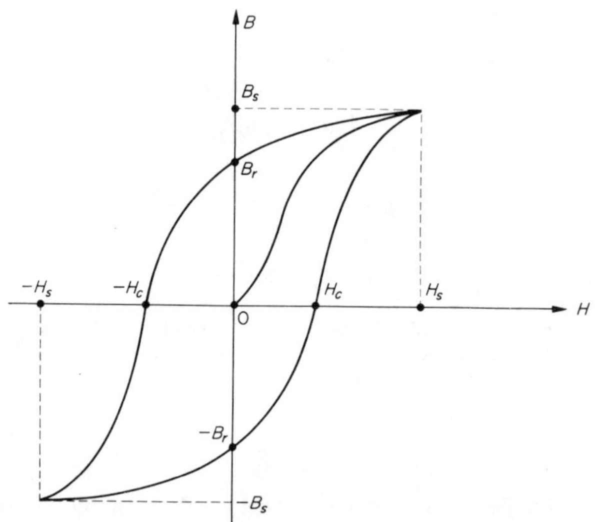

Fig. 9.2. Hystersis loop.

residual magnetization of the material is represented by the intersection of the return curve from B_s with the vertical axis at B_r. Changing the direction of the magnetizing force and repetition of the process will result in establishing a complete symmetrical loop about the point zero. This hysteresis loop which encloses the largest area is of particular significance and is often called *the* hysteresis loop of the material. The value of H_c is known as the coercive force and the ratio of B_r to B_s as the squareness ratio.

The hysteresis loss per cycle per unit volume is equal to the area of the hysteresis loop. Magnetically soft materials such as the ones used in magnetic heads have a narrow hysteresis loop, i.e. small coercive force whereas magnet-ically hard materials require large values of H for magnetization and demagnet-ization. Magnetic materials which are good electrical conductors exhibit high eddy current losses; hence, the heads are divided into thin, preferably insulated laminations.

Figure 9.2 clearly indicates the dependence of the permeability μ of the material on the magnetizing force H. The initial permeability μ_i is defined as the slope of the virgin magnetization at the origin (Fig. 9.3). The total permeability at point P is

$$\mu = \frac{B}{H}, \tag{9.17}$$

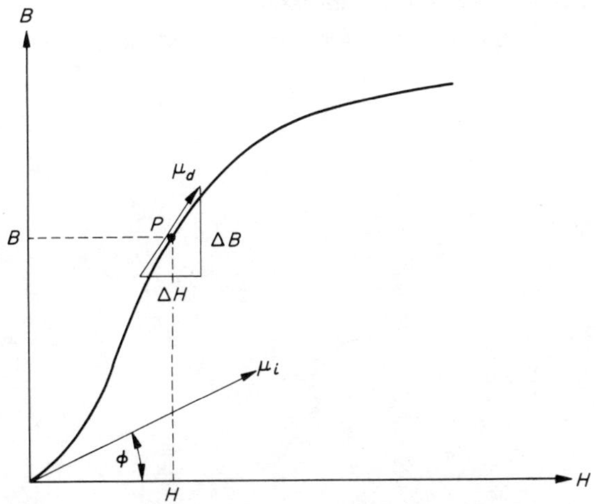

Fig. 9.3. Virgin magnetization curve. Initial permeability $= \tan^{-1}\varphi$.

and the differential permeability

$$\mu_{\mathrm{diff}} = \frac{\mathrm{d}B}{\mathrm{d}H}. \tag{9.18}$$

Finally the incremental permeability is defined as the permeability obtained by small variation of the magnetizing force around a constant (DC) magnetizing field. The maximum permeability, $\mu_{\max}$ may be considerably higher than the μ_i initial permeability.

Magnetic circuits can be treated in the same manner as electrical circuits by introducing the magnetic equivalent of current, voltage and resistance. Equating flux ϕ with current I, the magnetomotive force $Hl = nI$ with the electrical potential difference and introducing reluctance as equivalent for resistance the equivalent of Ohm's Law can be written as

$$R = \frac{nI}{\phi} = \frac{Hl}{\phi} = \frac{l}{\mu A} = \frac{n^2}{L},$$

(9.19)

for closed magnetic path. Here n is the number of turns, l the length of the magnetic path, L inductance, A the area cross-section and $\mu = \mu_0 \mu_r$ the permeability of the core material. Reluctances can by thought connected in series or in parallel and manipulated as resistances.

REFERENCES

1. Rayner, G. H. and Drake, A. E. (1970), *SI units in electricity and magnetism*, National Physical Laboratories.
2. Young, L. (1969), *Systems of units in electricity and magnetism*, Oliver & Boyd.

10 Standards relevant to digital magnetic tape recording.

BRITISH STANDARDS

BS 3968	7-track magnetic tape for data interchange recorded at 200 rpi.
BS 4503	9-track magnetic tape for data interchange.
	Part 1. Tape recorded at 800 rpi.
	Part 2. Tape recorded at 1600 rows per inch, phase encoded.
BS 4783	The care and transportation of magnetic tape.
BS 4732	Magnetic tape labelling and file structure for data interchange.

ANSI STANDARDS AND DRAFTS

X3B1/553-R	Proposed revision of X3.14, recorded magnetic tape for information interchange, 200 CPI, NRZ1.
X3B1/554-R	Proposed revision of X3.22, recorded magnetic tape for information interchange (800 CPI, NRZ1).
X3B1/555-R	Recorded magnetic tape for information interchange (1600 CPI, PE)
X3B1/556-R	Unrecorded magnetic tape for information interchange (9-track 200 and 800 CPI, NRZ1 and 1600 CPI, PE).
X3.27	Magnetic tape labels for information interchange.
X3B1/579	Magnetic tape cassette for information interchange.

ECMA STANDARDS

ECMA-5 – Data interchange on 7-track magnetic tape, 3rd edition (June 1970).

ECMA-12 – Data interchange on 9-track magnetic tape at 32 bits per mm (800 bpi), 2nd edition (June 1970).

ECMA-13 – Magnetic tape labelling (Nov. 1967).

ECMA-34 – Data interchange on 3.81 mm magnetic tape cassette (32 bpmm, phase-encoded) (Sept. 1971).

ECMA-36 – Data interchange on 9-track magnetic tape at 63 bpmm (1600 bpi) phase-encoded (Dec. 1971).

11 List of principal symbols

α, β, γ constants
κ, K susceptibility
η head efficiency
ρ density
ρ^* volume charge distribution
μ permeability; viscosity of air; friction coefficient
ϕ magnetic flux
σ_s saturation moment per gram
λ wavelength on tape
Θ wrap-angle
ω angular frequency

A cross-section area
B magnetic induction, magnetic flux density
C constant
CRC   Cyclic Redundancy Check
$E(x)$ error pattern
H magnetic field strength
I, J current; intensity of magnetization
K constant
L inductance
M magnetization; magnetic moment per unit volume
R reluctance; head radius
T tape tension

a transition length parameter
b thickness of backing material
d head-to-medium separation
e voltage
f frequency
g gap length
h air film thickness; head pole/tip depth; distance

i	current
k	constant
l	length
n	number of turns
r	radius, distance
s	distance
t_m	medium thickness
v	velocity
w	track width

Subject index

Author index